NASA SP–4311

WALLOPS STATION AND THE CREATION OF AN AMERICAN SPACE PROGRAM

Harold D. Wallace Jr.

The NASA History Series

National Aeronautics and Space Administration
NASA History Office
Office of Policy and Plans
Washington, D.C.   1997

Library of Congress Cataloging-in-Publication Data

Wallace, Harold D., 1960–
    Wallops Station and the Creation of an American Space Program /
Harold D. Wallace Jr.
      p. cm.— (The NASA history series) (NASA SP: 4311)

    Includes bibliographical references (p.) and Index.
    1. Wallops Flight Facility—History. 2. Astronautics—United
States—History. I. Title. II. Series: NASA SP: 4311.
TL862.W35W35   1997   97-30983
629.4'09755' 16—dc21   CIP

To the Memory of Florence C. Anderson—
who always believed that an education was something that
could never be taken away.

TABLE OF CONTENTS

ACKNOWLEDGMENT

This thesis marks the culmination of my master's program at the University of Maryland, Baltimore County, and I owe thanks to a number of people for its successful completion.

I wish first to thank the mentor of this work, Joseph N. Tatarewicz, for his encouragement, advice, and constructive criticism. His optimism and support kept this endeavor on track, and his willingness to work with my sometimes inconvenient schedule was greatly appreciated.

Also deserving of thanks are the members of my review committee in the History Department of UMBC: Sandra Herbert, Joseph L. Arnold, and Gary L. Browne, all of whom waded through the review copy of this work on short notice, and provided pertinent and instructive comments.

The research for this work would have been vastly more difficult if not for the assistance of Keith Koehler, Public Information Officer at NASA's Wallops Flight Facility. His guidance regarding Wallops' record collection, and his assistance in arranging interviews was indispensable. I also wish to recognize the aid rendered by Roger D. Launius and Lee D. Saegesser at the NASA History Office in Washington. Their comments and counsel helped me to maximize scarce research time. Similarly, Richard T. Layman and Garland Gouger at NASA's Langley Research Center kindly took time from other duties to facilitate my research during my visit there.

I would especially like to thank Robert T. Duffy, J. Chris Floyd, Marvin W. McGoogan, Joyce B. Milliner, Joseph E. Robbins, and Abraham D. Spinak, all of whom graciously consented to provide the oral history that was needed to fill in the gaps in the written record.

Finally, I thank my family, friends, and co-workers for their understanding and support during this project. Their patience during those times I was consumed with this work is greatly appreciated. Thanks particularly to Catherine Anderson for help with the tables, and Michelle Wallace for the map and chart. Responsibility for errors of fact or interpretation, of course, rests solely with the author.

ABOUT THE AUTHOR

Harold D. Wallace Jr. is an historian working in the Electricity and Modern Physics Division of the National Museum of American History. Current projects include an exhibit studying invention in modern lighting and research into historic aspects of restructuring in the electric power industry.

A native of Baltimore, Maryland, Mr. Wallace received an M.A. in the history of technology from the University of Maryland, Baltimore County. Prior to employment with the Smithsonian, he enjoyed a successful management career in the retail and wholesale hardware trades. He is a member of the American Historical Association, the Baltimore County Historical Society, and the Society for the History of Technology.

LIST OF ACRONYMS

ABMA: Army Ballistic Missile Agency
AEC: Atomic Energy Commission
AFRS: Auxiliary Flight Research Station
AGARD: Advisory Group for Aeronautical Research
AMPD: Advanced Materials and Physics Division
AMR: Atlantic Missile Range (Cape Canaveral, Patrick AF Base)
AO: Administrative Operations
ARC: Ames Research Center
ARDC: Air Research and Development Command
ARPA: Advanced Research Projects Agency
CAA: Civil Aeronautics Administration
CNAS: Chincoteague Naval Air Station
CNO: Chief of Naval Operations
CoF: Construction of Facilities
DOD: Department of Defense
DOVAP: Doppler Velocity and Position
FAA: Federal Aviation Administration
FRC: Flight Research Center (HSFS)
GEOS: Geodynamics Experimental Ocean Satellite
GSFC: Goddard Space Flight Center (Beltsville Space Flight Center)
HSFS: High Speed Flight Station (FRC)
IBM: International Business Machines
ICBM: Intercontinental Ballistic Missile
IGY: International Geophysical Year
IQSY: International Year of the Quiet Sun
IRD: Instrument Research Division
JPL: Jet Propulsion Laboratory
LaRC: Langley Research Center (LMAL, LAL, LRC)
LeRC: Lewis Research Center
MIT: Massachusetts Institute of Technology
MSC: Manned Spacecraft Center (Houston)
MSFC: Marshall Space Flight Center
MSFN: Manned Space Flight Network
MSTS: Military Sea Transport Service
NACA: National Advisory Committee for Aeronautics
NASA: National Aeronautics and Space Administration
NATO: North Atlantic Treaty Organization
NBS: National Bureau of Standards
NRL: Naval Research Laboratory
OLVP: Office of Launch Vehicle Programs
OSFD: Office of Space Flight Development

OSFP:	Office of Space Flight Programs
OSO:	Orbiting Solar Observatory
OSSA:	Office of Space Science and Applications
OTDA:	Office of Tracking and Data Acquisition
PARD:	Pilotless Aircraft Research Division
PMR:	Pacific Missile Range (Vandenberg AF Base)
PSAC:	President's Science Advisory Committee
RAM:	Radio Attenuation Measurement
R&D:	Research and Development
RAND:	Research And Development Corporation
RCA:	Radio Corporation of America
S&E:	Salary and Expenses
Scout:	Solid Controlled Orbital Utility Test System
Spandar:	Space Ranging Radar
SRB:	Solid Rocket Booster
SSUA:	Special Subcommittee on the Upper Atmosphere
STADAN:	Space Tracking and Data Acquisition Network
STG:	Space Task Group
TAGIU:	Tracking and Ground Instrumentation Unit
Tiros:	Television Infra-Red Observation Satellite
TOS:	Tiros Operational System
UARRP:	Upper Atmosphere Rocket Research Panel
USNS:	United States Naval Ship
VPI:	Virginia Polytechnic Institute

MAPS OF WALLOPS

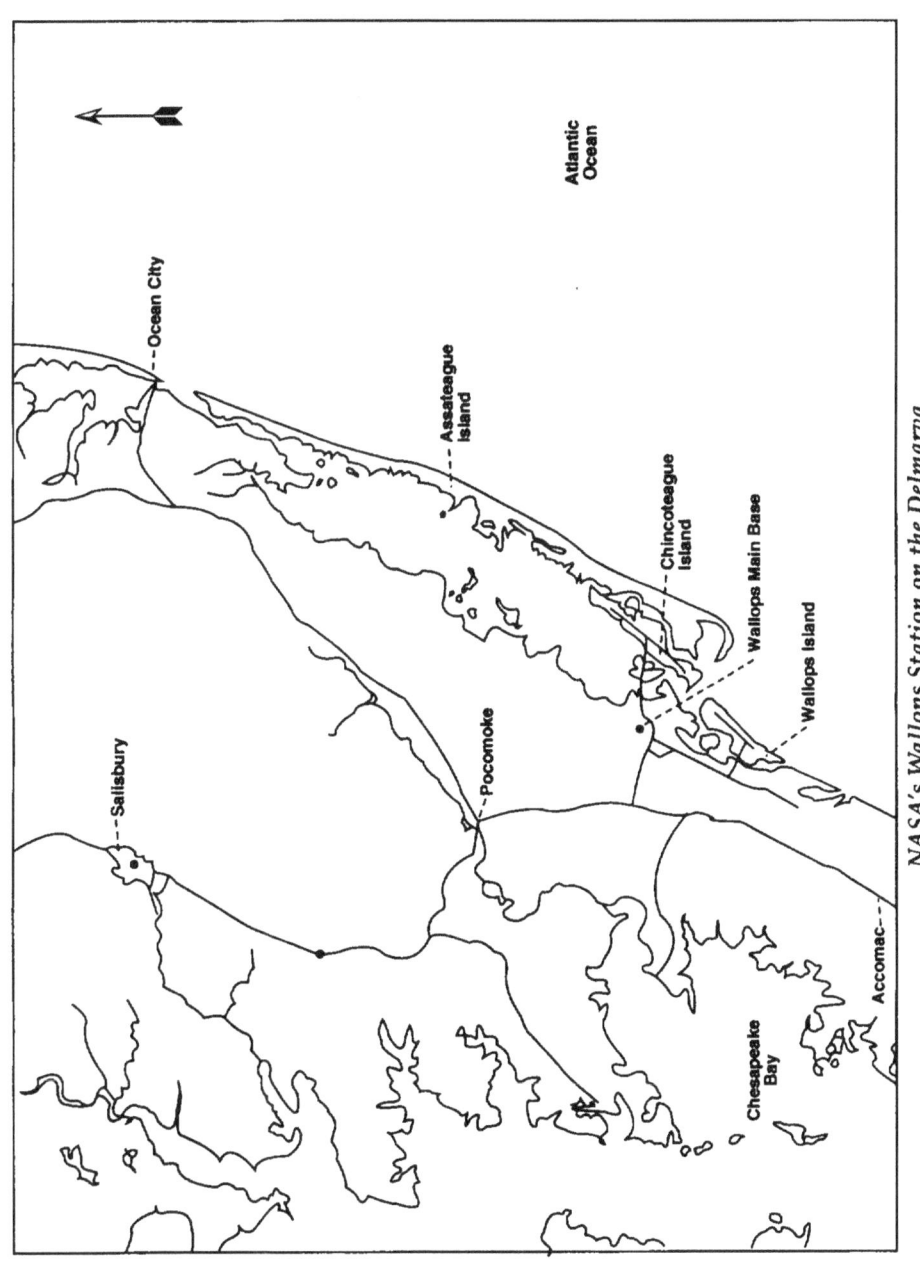

NASA's Wallops Station on the Delmarva

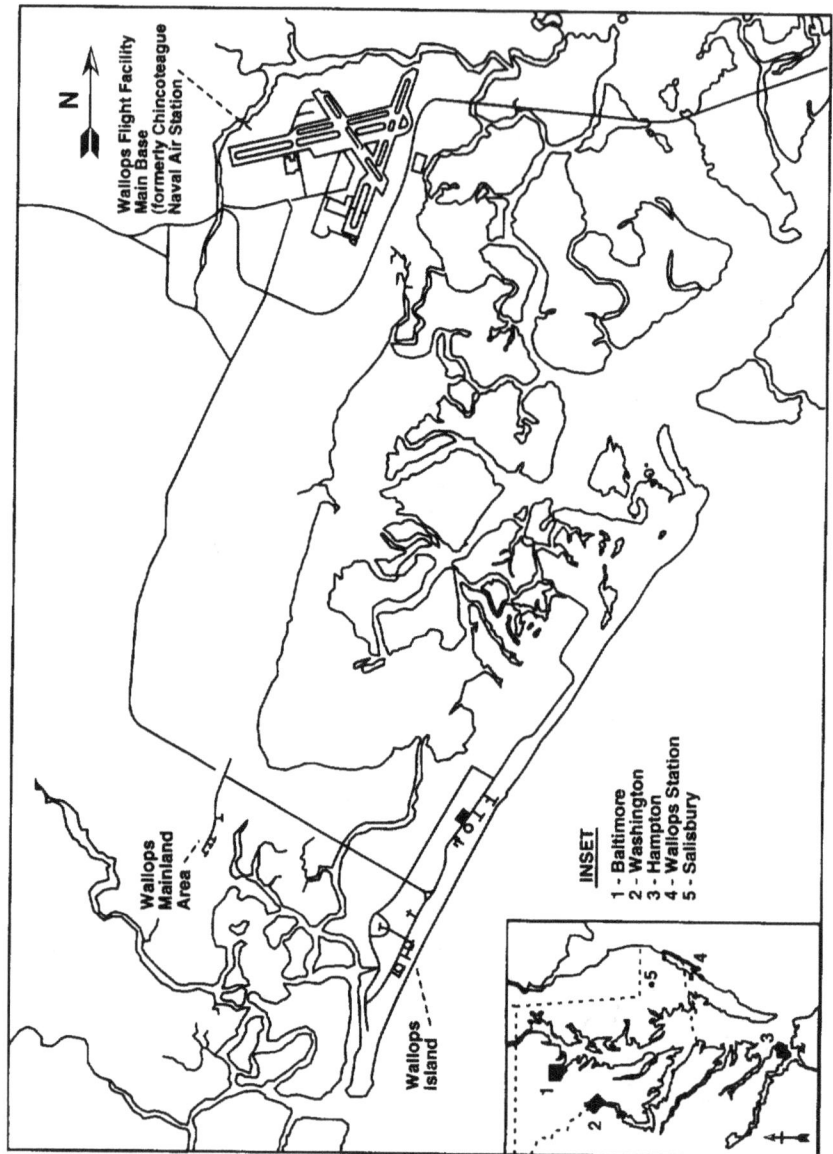

N

Wallops Flight Facility
Main Base
(formerly Chincoteague
Naval Air Station)

Wallops
Mainland
Area

Wallops
Island

INSET
1 - Baltimore
2 - Washington
3 - Hampton
4 - Wallops Station
5 - Salisbury

NASA's Wallops Station (in three sections)

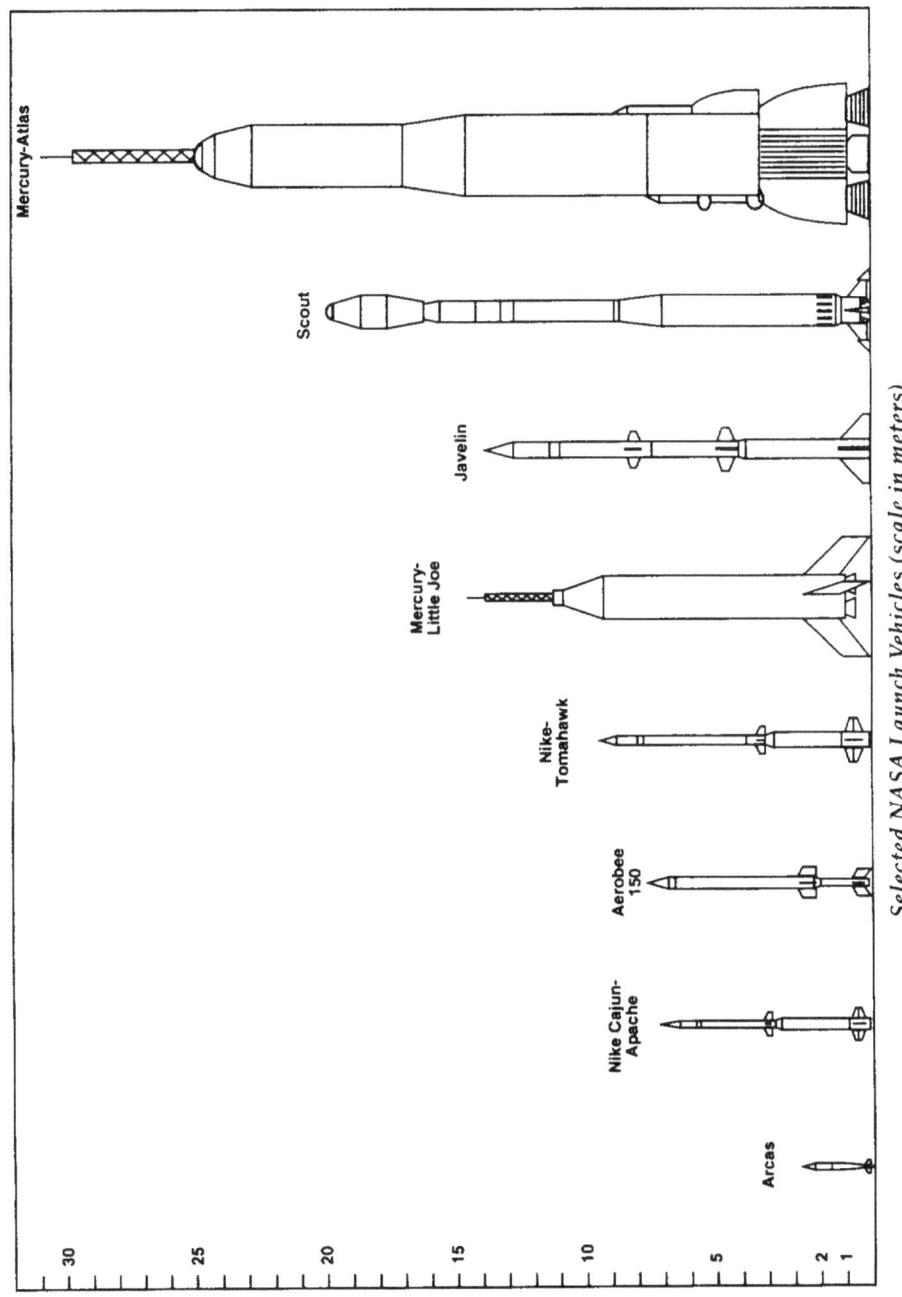

Selected NASA Launch Vehicles (scale in meters)

Chapter 1

INTRODUCTION

The course of the Space Age underwent a fundamental shift during the decade of the 1980's. The heady era of Sputnik, Apollo, and the Cold War-fueled space race shifted to an era of more methodical activities as space operations became popularly mundane. Similarly, seminal works pertaining to the history of conspicuous early space projects have been joined on the library shelf by words examining less glamorous, but still important topics. The big, visible space projects existed as much for reasons of politics and national prestige as for scientific research, and thus attracted the early attention of historians. The advancement of human knowledge and skill, however, owed at least as much and perhaps more to smaller projects and research conducted out of the spotlight.

A facility devoted to such projects still operates on the East Coast of the lower Delmarva Peninsula. The National Aeronautics and Space Administration's Wallops Flight Facility has, since its establishment in 1945, launched over 14,000 rockets, making it one of the most prolific launch sties in the world as well as one of the least known.[1] Currently a subsidiary of Goddard Space Flight Center, Wallops has always been among the smaller of America's aerospace research facilities. Despite, and possibly because of, the administrative and budgetary chaos that often characterized this nation's space effort, Wallops evolved from a highly specialized, test facility to a more generalized, multi-faceted research center. The history of the base reaches back to the early days of U.S. involvement in space research, and reflects most of the major controversies encountered therein. An historical examination of the base, therefore, has value not only because of the comparative lack of such attention, but also because it allows a unique vantage point from which to view what is, to paraphrase policy historian John Logsdon, "the great adventure of our lifetime."[2]

Due to the limitations of size and time inherent in a thesis, this work will not be a detailed, all-encompassing history of the Wallops Station. This thesis will focus on the political, administrative, and social history aspects of the base from 1957 to 1966. This period began with the launch of the Soviet Union's *Sputnik 1*, continued through the creation of NASA, and culminated at the height of the Project Apollo escalation. A fast-paced era, it also includes the second of the three most important periods in Wallops history to date.[3]

The thesis is arranged in five chapters, with this first serving to provide background information on the base, relate events leading up to the creation of NASA, and introduce most of the themes that run through the body of the work. The second chapter discusses the founding of NASA, the subsequent expansion of Wallops, and the organization of the base as an independent administrative entity. During this immediate post-Sputnik era the differing prioritization of aeronautical and space science research within NASA began to fundamentally alter Wallops' mission. The sudden appearance of a space race, combined with the unexpected closure of a nearby military base, also shifted the relationship between the station and the local community.

Chapter three provides a look at Wallops' involvement with the U.S. piloted space flight effort. This involvement, heavy during Project Mercury, declined throughout the period until almost nil during Project Apollo. The staff's reaction to the novelty of press coverage and public interest in its operations, a side effect of the piloted programs, is also examined.

Chapter four traces the course of space science research at Wallops by discussing not only programs and facilities located at the Virginia base, but also those operations that occurred off-range at various locations. Wallops' significant role in NASA's program of international cooperation coalesced during this early period, and is also examined.

The final chapter explores how the period of relative stability at Wallops through the following decade, extended from changes (and non-changes) that occurred during the transition era. The roles of Wallops' various customers are summarized, as is the role of the Station within both the local environment and within NASA. Before launching into the account of such an active era, however, it would be well to set the stage.

Langley Aeronautical Laboratory originally established the facility at Wallops Island in 1945 to fulfill an urgent wartime requirement for a test range to provide militarily vital aeronautical engineering data. Today the base serves scientists as the nation's only civilian controlled launch range supporting a wide assortment of research projects; a radical change. Conversely, the primary method utilized by the researchers at Wallops, the launching of solid-fueled rockets, remains little changed from the early days. Defined, founded, and operated by Langley engineers, the island base initially reflected its parent lab in many ways. Thus, a review of the backgrounds of both Langley and the National Advisory Committee for Aeronautics (NACA), of which it was a part, becomes necessary. This account begins in the opening years of the twentieth century, at the dawn of heavier-than-air flight.

Aviation existed in the embryonic state until the onset of this century. The lighter-than-air craft of the late nineteenth century exhibited little improvement over that flown by the Montgolfier brothers in 1783. Advancement came slowly until the success of the Wright brothers in 1903. The early years of powered flight found the pioneers of aviation struggling to understand both the physical properties of the atmosphere and the basics of aeronautical engineering. The experience of the First World War demonstrated that European researchers had advanced more quickly in comprehending and utilizing the new arena than had their American counterparts. The miliary, and to a lesser extent commercial, implications of this American aeronautical deficiency prompted the U.S. government to take action.

In March of 1915 Congress passed, appended to a naval appropriations bill, a law establishing the National Advisory Committee for Aeronautics. The somewhat general wording of the law empowered the Committee, "to supervise and direct the scientific study of the problems of flight with a view toward their practical solution..."[4] Established despite misgivings from the military (worried about a civilian agency syphoning off resources), and bureaucratic squabbling common in Washington, the Committee met for the first time on 23 April 1915 in the offices of the Secretary of War. Though ostensibly civilian in nature, five of the original twelve seats on the Main Committee were held by military aviation personnel. This set a pattern for the special relationship that existed between the NACA and the services. No matter how busy, the Committee remained responsive to the needs of this prime customer throughout its existence.[5]

Once organized, the first priority of the Committee was the construction of a research laboratory. They believed modern facilities and motivated personnel would give them the ability to compete with Europeans who owed much of their technological lead to such state sponsored concerns. The War Department, already directed by Congress to select a site for such a facility, recommended a site near the town of Hampton, Virginia. The NACA concurred with this choice which offered reasonable proximity to Washington headquarters and Virginia industry, a variety of "experimental flying conditions," the promise of an adjacent military airfield, and enough isolation to ensure both safety and security.[6] Langley Memorial Aeronautical Laboratory, dedicated in 1920, became the foundation of the NACA, and profoundly influenced national and international research for decades to come.

The early days at Langley were far from comfortable. Located in the midst of farmland just off the Chesapeake Bay, conditions bordered on the primitive. Scarce housing, an isolated location, and a disagreeable climate prompted more than a few resignations at the beginning.[7] As time passed and conditions slowly improved, however, a formidable research institution grew. Some of the world's most advanced wind tunnels and test equipment went into operation despite the lean budget years of the Great Depression. This allowed the engineers at the lab to do pathfinding work in aeronautical engineering. It should be noted that "engineering" and "science" do not always mean the same thing. The Langley engineers concentrated on designing and improving flying equipment rather than attempting to conduct research into atmospheric phenomena for its own sake. This focus on inventing and refining hardware was, of course, their job, but it would lead to problems later on.[8]

A relaxed, scholarly atmosphere prevailed at the lab, informality being viewed as a stimulus to creativity. The comment made by a senior engineer, "Let's try the damn thing and see if we can make it work," illustrated a true understanding of the nature of experimentation, which required a tolerance for occasional failures.[9] Situated at a distance from the NACA's Washington headquarters sufficient to escape stifling managerial scrutiny, Langley prospered and came to regard its relative independence as a fundamental necessity. Most research work was performed "in-house" rather than contracted out, and many Hampton residents found jobs at the lab. As time passed, the lab and the community adjusted to each other.

The advent of World War II did not take the NACA completely by surprise. Indications of advancing German aeronautical research abounded for several years prior to American military involvement. In an effort to accelerate the pace of U.S. research, the NACA persuaded Congress to authorize two new laboratories. Ames Laboratory, in Sunnyvale, California, opened in 1940 and provided testing facilities close to the West Coast aircraft manufacturers. Less than a year later, Lewis Laboratory, built in Cleveland. Ohio, began providing data on aircraft engines.[10] In both cases Langley personnel were dispatched to plan, oversee, and operate the new labs. Thus, the Langley methodology spread through the growing NACA field organization. This methodology combined a commitment to the research ethic (intellectual freedom and systematic procedures), a certain level of administrative independence, and an aversion to contractors. While promoting the desired research standards, this methodology also promoted its share of tension between the field personnel and a Headquarters staff trying to maintain control over an expanding organization.[11]

The war served to intensify the military's claim of pre-eminent access to the NACA's facilities. From the beginning, the Committee gave specific military projects priority over the general research it preferred doing.[12] The "clean-up" work performed during the war contributed greatly to the allied victory as almost every U.S. combat aircraft-type flown spent some time in an NACA wind tunnel. NACA leaders found this developmental role distasteful, but the needs of a nation at war left them with little choice.[13] Aircraft manufacturers working on non-military projects found themselves unable to obtain similar services for their commercial designs. The NACA viewed its role as one of providing general data that all manufacturers could use. They worked carefully to avoid any charges of interfering with free-market competition, or allowing publicly funded facilities to assist private gain, while trying to be responsive to industrial needs.[14]

Two important fields of inquiry rose to the top on the research agenda during the years of the Second World War: high-speed flights, and missile development. Dealing with these topics required new techniques and new facilities, all planned and built with war-time haste. With Ames and Lewis still in the very early stages of operation, the burden of this research fell

4

largely on Langley. Aircraft speeds rose steadily throughout the 1920's and 30's as more powerful engines and more efficient designs came into service. The speed of sound became a tangible milestone. Goal to some, barrier to others, respected scientists and engineers argued about the possibility of exceeding Mach 1. Not subject to debate, however, were the real aerodynamic effects created by aircraft approaching this velocity. Air piled up in the front of a fast moving plane, causing severe buffeting and loss of control. These "compressibility effects" began to cause slips in manufacturing schedules and cost pilots their lives. Research into the transonic speed range became vitally necessary and the NACA began its research.[15]

Unfortunately, strange things happened in wind tunnels during tests at these speeds. Data readings, accurate above and below the transonic range, grew inaccurate within that range. A condition engineers referred to as "choking" occurred when shock waves generated by air moving over a test model rebounded off the tunnel walls, interacting with the model. These frustrating difficulties led NACA researchers to consider new methods of obtaining test data. One such method, designing and flying experimental research aircraft, led to the establishment of the High Speed Flight Station. Opened in 1946 adjacent to Edwards Air Force Base in California, this station gave Langley researchers a place to test-fly new designs, resulting in the famous X-series of aircraft. Two other methods, propelling instrumented models to high speeds by use of a rocket motor, and dropping instrumented devices from a high–flying aircraft, also required the establishment of a specialized facility.[16]

Concurrent with the need to conduct transonic flight research came the need to test early missile designs that began to appear late in the war. Several missile designs underwent testing in the Langley tunnels during the war, but there existed some question as to the status of this new device. Were they "pilotless aircraft" and subject to the NACA's research mandate, or ordnance, a glorified bullet, and out of the NACA's purview? Though solely a military device at the time, the NACA adopted the former position and started looking for a range from which they could test missile guidance and propulsion systems.[17]

In December 1944, Langley's Acting Engineer-in-Charge, John W. Crowley, organized a Special Flying Weapons Team to "oversee all missile research" at the lab. This team, led by Crowley himself, recommended the establishment of an Auxiliary Flight Research Station for the conduct of both high-speed flight and missile tests. The proposed base needed clear, unpopulated space downrange, a series of locations parallel to prospective flight paths suitable for radar tracking stations, and a reasonable proximity to Langley. Safety and security considerations dictated an isolated spot and a nearby military airfield was deemed a must.[18]

A site near Cherry Point, North Carolina, drew the attention of the Langley engineers. Launches could be directed out over the Atlantic with flight paths parallel to Cape Hatteras. Less than an hour away from Hampton by air, with a nearby Marine air base, this site seemed ideal. However, anticipated difficulties getting to tracking sites on the barrier islands combined with unanticipated objections to this civilian plan from the officers at the Marine base and eliminated Cherry Point from consideration.[19] Crowley's team then re-examined a site originally rejected as too remote: Wallops Island.

Home to an old Coast Guard station and owned by a group of Pennsylvania sportsmen, support facilities for both people and experiments left much to be desired. Yet the lure of a base near Langley, with a clear range out over the Atlantic, good locations south along the coast for tracking stations, and the adjacent Chincoteague Naval Air Station, proved irresistible. In April 1945 Congress appropriated funds for the research station, and an accompanying facility at Langley. Navy plans to use the north end of the island as an ordnance test site, which included missile launches, settled the matter. On 11 May, 1000 acres on the south end of the island were leased by the NACA, clearing the way for the hiring of employees and the shipment of materials.[20] Crowley pulled engineer Robert Gilruth out of Langley's Flight Research Division and put him in charge of the new organization.[21] Gilruth and his associates tackled the job of preparing the site for rocket operations, organizing the facility, and commencing launches.

The hectic pace of activities did not slow with the end of the war in Europe. Launch operations from hastily constructed temporary facilities on Wallops commenced on 27 June 1945. With no experience in the conduct of rocket operations, Langley relied on the assistance on the Navy's Bureau of Ordnance, their neighbors on the island, until their own personnel gained proficiency. Gilruth delegated the tasks of assuring that the Langley personnel achieved such proficiency to engineer William J. O'Sullivan; Gilruth himself was busily coordinating a variety of other tasks.[22]

Like the military facilities at Cape Canaveral, Florida, and White Sands in New Mexico, the civilian range at Wallops quickly became host for a number of research projects and capabilities promoting aeronautical research. The southern tip of the island served as a drop zone for free falling models. Though not utilized to the same extent as the rocket model method, drop models did provide useful data. Balloon launches relayed atmospheric data in support of flight operations, an important function since the atmosphere, unlike the environment inside a wind tunnel, could not be carefully controlled. A desire to experiment with ramjet designs led to the early construction of a wind tunnel facility known as the Preflight Jet, the only one of its kind at that time.[23]

The establishment of Wallops paralleled in many ways the establishment of Langley Lab. Engineers from the NACA came together in a remote location, supported by the military, assisted by local workers, to conduct pathfinding research into a highly technical enterprise of vital and urgent interest to a country at war. Early conditions at Wallops also recall the early days at Langley. The sparsely populated area contained little save farmland and

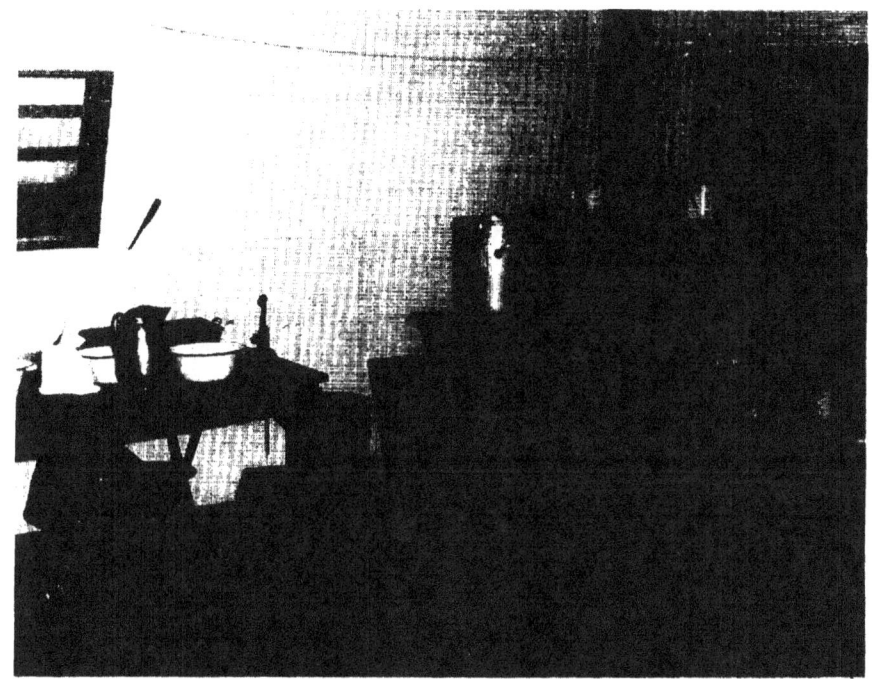

The kitchen wing of Wallops quonset hut hotel, August 20, 1945.

marshes, demanding a measure of endurance form those assigned there. Scarce housing, few social diversions, and a general lack of amenities made working there an unappealing prospect. The island itself was barren of facilities. No road connected it to the mainland, so reaching it required a ferry or seaplane. Portable generators provided power. Supplies as basic as water needed to be ferried in. Food prices in the area soared, and the nearest hospital facilities lay forty miles away in Salisbury, Maryland, as the naval base at Chincoteague could only provide emergency services. An abundance of mosquitoes and horseflies sufficed to round out a very uncomfortable duty station.[24]

The remoteness of the location served its purpose, however. It satisfied the engineers who wanted to conduct hazardous operations safely and without interruption, and pleased a military clientele concerned with maintaining a shroud of secrecy around an emerging class of weapons. The existence of the station was not publicly acknowledged for over a year, and the research results during the early period quite often were released only on a need to know basis.[25] The isolation also fostered the Langley traditions of a relaxed atmosphere and relative freedom from managerial scrutiny.

The rudimentary nature of the facilities at the Station began to change the day the war ended. The process of constructing a permanent plant commenced with the opening of bids at Langley. The post-war scale back in government spending slowed the construction process, as did the Navy's oft-stated plans to purchase the entire island for the Bureau of Ordnance. While the Navy professed no objection to sharing the island with the NACA, they delayed. By law the government agencies could only build on government owned land, and the Bureau of the Budget refused to allow the NACA to purchase parts of an island scheduled for acquisition by the Navy. Therefore, the temporary facilities first erected saw use for several years longer that anticipated, with permanent construction limited to the few acres already purchased.[26]

The administrative organization coalesced somewhat more quickly during this period. On 10 June 1946, the Auxiliary Flight Research Station became the Pilotless Aircraft Research Division of Langley Laboratory (PARD). Wallops remained the operational site of the group and officially received the name Pilotless Aircraft Research Station, though the acronym PARS never found widespread use and the base continued to be called simply: Wallops. Robert Gilruth, designated division chief, started to refine his team, a task largely completed by 1950.[27]

For Wallops, the most profound effect of this reorganization turned out to be the assignment of Robert L. Krieger to the post of Engineer-in-Charge, Wallops Island, a position he held until his retirement from NASA in 1980. A Hampton native, Krieger worked at Langley in his youth, performing various unskilled and semi-skilled tasks. He eventually found himself working for engineer Edmund C. Buckley in the Photo Lab. Buckley persuaded Krieger to seek an engineering degree, and after taking this advice Krieger graduated from the Georgia Institute of Technology in 1943. He returned to Langley and was assigned to the Instrument Research Division, now headed by Buckley, and proceeded to work on radar tracking and photographic data collection techniques. When Buckley accepted the position of Assistant Chief of PARD in 1948 he called upon his protege to take charge of operations at Wallops. Krieger's appointment served to highlight the importance of the tracking and data acquisition function of the PARD operation. Launching the rockets was only a part of the research process. New radar tracking, radio telemetry, and photographic techniques played an indispensable role in conducting a successful project, and Krieger had specialized in that area. He did not just bring technical experience and a background steeped in the Langley tradition to the base, he proceeded to make Wallops his own.[28]

Other personnel shifts of importance to Wallops occurred during this period. John Crowley moved to NACA Headquarters, becoming Assistant Director of Aeronautical Research. He worked closely with both Dr. George W. Lewis and Dr. Hugh L. Dryden, the NACA's last two research directors. When Crowley's transfer became permanent in July 1947 Floyd L. Thompson succeeded him as Langley's Research Department Chief with Ira A. Abbott and Special Flying Weapons Team veteran Hartley A. Soule as his assistants.

Within a year, Abbott transferred to Headquarters to assist Crowley.[29] It is apparent that even though the NACA had, by 1950, grown into an organization far larger than any of its founders could have foreseen, upper management remained a tight little group. They knew each other, shared common backgrounds, and knew intimately how the NACA functioned. Most of them also knew Wallops, an important consideration given their prominent role in the creation of NASA.

During the first phase of operations at Wallops, the transonic period, the predominant number of tests fell into the category of basic research. This included launches to investigate drag, control, and stability characteristics of assorted generic aerodynamic shapes. Interspersed with these general tests were examinations of specific military aircraft and missile models, but the relative numbers indicate the weight Langley engineers gave research over development.[30] (See appendix 1)

Regardless of the manner of the test or the customer, the overall frequency of testing rose steadily. The value of the data generated at Wallops prompted an industry request in 1948 for the NACA to expand and accelerate the PARD program. The simultaneous growth of operations at the Naval Air Ordnance Test Station, on the other end of the island, caused concerns about potential range interference. The issue came to a head in late 1948 and early 1949, and resulted in the Navy acceding the "primary interest" of NACA activities on the island. The establishment of test ranges at Point Mugu and Point Arguello, California, lowered the Navy's interest in Wallops and cleared the way for NACA purchase of the island.[31] By use of condemnation proceedings the government took possession of the island on 7 November 1949, and later paid $93,238.71 in compensation to the previous owners. This finally allowed the needed construction to proceed.[32]

This construction centered mostly on replacing the old temporary structures, and erecting the shops and control facilities needed to handle an increasing workload. A test apparatus known as a helium gun was transferred from Langley to Wallops at this time, adding to the research arsenal at the base. The number of employees assigned to the station stabilized at around 75 during this period, however, a situation which did not change appreciably until the creation of NASA. The internal organization of the base also stabilized with the assignment of personnel to either the Mechanical Services Unit, the Research Section, or the Administrative Unit. "In the daily operations at the island, however, organizational lines were not rigidly drawn —all personnel helped in any way they could to get the job done".[33]

The rough local conditions continued to stress those at the base. The scarcity of community facilities caused hardships for all, even those at the top. In 1951 Robert Krieger requested permission to move his office back to Langley. He retained his position in the Wallops structure with little effect on the operations at the base since planning, budgeting, data reduction, and many of the test preparations took place at the Hampton lab. Engineer John C. Palmer assumed responsibility for overseeing the daily operations on the base. This would solidify the administrative pattern that would prevail for the rest of the decade. Researchers, inside or outside of the NACA, who wished to use the facilities at Wallops went to PARD at Langley, because the decision-making process operated there. Wallops performed the same function as any of Langley's wind tunnels, researchers traveled to the Station only to conduct their tests. Wallops provided services, PARD provided direction, and Langley provided support.[34]

The focus of the research performed at Wallops began to shift in the early 1950's. Transonic research remained important for several more years, but equipment designers discovered a way around the choking problem. The new fixtures, slotted-throat wind tunnels, could provide transonic data without the necessity of watching an expensive, highly instrumented model vanish into the ocean. A major source of contention between advocates and opponents of rocket model testing lay in the waste inherent in the method. Models required money and, more importantly, time to produce and outfit with equipment. Proponents justified the tests by pointing out that they lacked a less-expensive method. The slotted-throat tunnels removed this argument, which was one of the major reasons for Wallops' existence. In addition to the quality of tunnels, the quality of equipment capable of conducting transonic research at Ames and Langley made the rocket model technique less necessary. At this point however, another research program arose to supplant transonic research in importance. Despite the execution of specific projects for various customers, the NACA felt that their primary mission remained basic, fundamental inquiry into the unexplored areas of flight. So, PARD found other uses for the Wallops range.[35]

Since the vivid German demonstration of the ballistic missile's military potential late in the war, American planners had slowly begun to investigate this weapon. The revolution represented by combining atomic bombs and pilotless aircraft started generating concern, especially after the Soviets broke America's nuclear monopoly in September 1949. The explosive force of an atomic bomb could compensate for the inaccuracies of early missile designs to a point. As the distance from the target increased however, the inaccuracies became unacceptable. The Communist victory in China, increasing tensions in Europe, and a war in Korea, all spurred U.S. missile research to overcome the technical difficulties.

By 1950 Wallops was conducting tests of the sub-sonic Snark cruise missile and its supersonic follow-on, the Boojum.[36] Cruise missiles, even if supersonic, suffered from the same vulnerabilities that endangered all combat aircraft. Again the Germans provided the lesson. The Allies shot down V–1 cruise missiles in droves, but could not devise a defense against the ballistic V–2 missile once it took off. Ballistic missiles presented much more complex problems by operating at higher speeds, reaching greater altitudes, and experiencing more heating than any weapon system previously designed. The research questions concerning drag, stability, control, and performance were old ones, while the territory being opened, the hypersonic speed regime, was virtually unexplored.[37]

The nature of the NACA's hypersonic program grew from its response to the earlier transonic problem. Again, wind tunnels could provide little data, requiring an alternate approach. Researchers at Ames and the High Speed Flight Station proposed extending the experimental aircraft program, while Langley and Lewis advocated increasing Wallops capabilities. NACA Headquarters, under political fire for having failed to exploit such advances as rocketry, swept wings, and jet engines during the war, needed a program that would put it back on the forefront of research. Transonic and supersonic research had been first steps. A hypersonic program provided the next logical step, especially since it coincided with emerging military needs.[38]

The PARD's experience at Wallops put the NACA in a very good position. Deriving accurate data using the rocket model technique required some expertise. One Ames engineer noted, "Most of the missile manufacturers are engaged in obtaining aerodynamic data from firings of their missiles. Almost without exception they would like to know the secret of PARD's success in getting reliable data from such firings."[39] After a 24 June 1952 meeting at Wallops, the NACA Committee on Aerodynamics adopted a resolution calling on the NACA to, "increase its program dealing with problems of unmanned and manned flight in the upper stratosphere at altitudes between 12 and 50 miles, and at Mach numbers between 4 and 10," and to, "devote a modest effort to problems associated with unmanned and manned flights at altitudes from 50 miles to infinity, and at speeds from Mach number 10 to the velocity of escape from the Earth's gravity."[40] The NACA Executive Committee adopted this position the following month. Desiring to take no chances, Headquarters elected to pursue a balanced program. One path led to the X–15 and ultimately the Space Shuttle, the other led to Projects Mercury and Apollo.[41]

Despite the completion of construction on the island in 1952, the redirected research effort called for new equipment and facilities. The boosters in use at that time could not easily reach hypersonic speeds. Bigger, more powerful boosters required larger launching equipment and more spacious shops. Also, better tracking and data acquisition hardware capable of supporting higher speeds and altitudes were essential. The final stumbling block concerned limitations placed on Wallops' range clearance. Even though the Navy curtailed operations on the island, the general area remained busy. Fleet training areas lay offshore, civil air routes passed nearby, and both Navy and Air Force conducted supersonic flight training over the ocean. Electronic, as well as physical interference posed difficulties. With a plethora of land based, shipboard, and airborne radars and radios close at hand, Wallops began to seem much less isolated.

The increasing traffic posed little problem early on as test rockets did not fly too high and ended their flights only a few miles offshore. The hypersonic program, with a reliance on multi-stage rockets, dictated sea and air clearances to much greater distances.[42] Early attempts to extend the sea range met with the determined opposition of the Navy's Commander-in-Chief, Atlantic Fleet. After several years of negotiations, the Navy and the NACA reached agreement on coordination of activities around Wallops Island. Although it occasionally needed fine tuning, this coordination worked well for the rest of the decade. Fortunately, cooperative use of military training areas meant that Wallops only rarely conflicted with civil air routes at this time.[43]

Though still somewhat spartan, the living conditions around the area began to improve. Part of the completed construction program included a building that functioned as cafeteria, lounge, and bunkhouse for both varieties of personnel: residents and transients. The permanent employees who operated and maintained the Station found housing in the local area, if indeed they were not already living there. The transient personnel, who came to the base only to participate in tests, usually stayed in the service building on the island. Social activities, aside from fishing, remained hard to find for these visitors, and many of them spent their spare time working on their projects.[44]

By mid–1953, projects, especially those from the military started running into serious delays. Attempting to increase the workload without significantly increasing the workforce, something Congress refused to allow, partially accounted for the growing backlog. The NACA's refusal to allow Ames to set up a Wallops-like facility on the West Coast added to the problem. The intricate and time consuming process of model preparation, the envied "secret" of PARD's success, completed the morass.[45] Langley engineers preferred to build their models in Langley's own shops, as industry-supplied models frequently failed to meet flight standards. Similarly, the lab's Instrument Research Division (IRD) had "hand-tailored" telemetry systems to the point where nothing available outside the lab gave suitable results. PARD adopted a policy of returning to the manufacturer models needing redesign or corrective work, and IRD gave the military models priority over the ones devoted to general research. It took over a year to get the test schedule back on track, and a tight pace remained the norm at Wallops.[46]

In late 1953 the NACA commenced a new program at the Station in response to a military request. The survivability of aircraft subjected to severe and sudden wind gusts, like those produced by an atomic blast, constituted an unknown factor in designing new aircraft. Langley developed a method of simulating such blasts by means of conventional explosives and measuring the aerodynamic loads exerted on scale models placed nearby to provide the needed data. After determining that accurate testing could not be safely conducted indoors at Langley, researchers began testing outdoors at Wallops. The engineers exploded charges as large as 650 pounds in the course of this

"Blast Research Project," which provided increased experience with both models and explosives at the station.[47]

The higher performance motors required for the hypersonic program started tests from Wallops in 1954. The Deacon rocket, produced by the Allegheny Ballistics Laboratory specifically for use as a research rocket was the booster of choice during Wallops' transonic phase. Fired singly or in clustered groups the Deacon remained a valuable tool for many years, but military Nike and Honest John rockets propelled the redirected research program. By late 1954, a four stage vehicle utilizing the Nike reached Mach 10, a significant advance. Engineers placed the speed of a re-entering intercontinental ballistic missile (ICBM) warhead at Mach 20, however, which called for more rocket power to meet test objectives.[48]

The Honest John promised to provide the desired performance but the destruction of a launcher during the first firing of the bigger booster emphasized the need for new launch equipment. New support facilities were also required. A second round of construction began which gave Wallops the ability to carry out its part of the hypersonic program. The local economy also derived some benefit from the expansion. While NACA personnel built the new launcher, for example, a local firm received the contract for the concrete pad. The new pad went into operation by the end of 1955. After a number of component tests, a five stage vehicle, the first ever launched, flew on 26 August 1956. PARD researchers calculated that this vehicle attained a speed of Mach 17.[49]

As Wallops gained experience with its new boosters, the engineers began obtaining valuable data on the flight characteristics of objects moving at re-entry velocities. Raw speed, of course, was not the end goal of the program. Dealing with the heat generated by objects traveling at such speeds provided the main impetus for the researchers. The quest to understand such a phenomena involved more that just rocket flights. The Pre-flight Jet Facility, modified at this time, allowed the Wallops engineers to conduct high-temperature testing in a laboratory setting. They tested a wide variety of shapes and materials at a range of temperatures and pressures. While not capable of producing the extremes found in actual flight, the tests of nose cones, fins, and scale models in the new Ethylene Jet refined the rocket model process. This combination of wind tunnel and rocketry put PARD and Wallops in the fore of hypersonic research.[50] Indeed, the high-temperature research became so important that the PARD altered its internal organization to promote the efficiency of the work and reflect the changing program. A High-Temperature Branch replaced the old General Aerodynamics Branch with engineer Paul Purser as its head.[51]

The mid-fifties also saw another novel type of research come to Wallops Station. Amid the flights of military models and general aerodynamics vehicles, scientific sounding rockets began to rise from the ocean-front launchpads.

After the war, a group of U.S. scientists formed the V–2 Upper Atmosphere Research Panel in order to probe the atmosphere with captured V–2 missiles. The number of V–2s being limited, the group turned to other vehicles to carry their instruments, and changed their name to the Upper Atmosphere Rocket Research Panel (UARRP).[52] In June 1947, Assistant to the Chief, PARD, William J. O'Sullivan became the NACA representative to the UARRP. In December, O'Sullivan also became the NACA member of the (NACA) Aerodynamics Committee's Special Subcommittee on the Upper Atmosphere (SSUA). This dual membership allowed O'Sullivan to keep both groups abreast of the current state of hypersonic research.[53] Dr. James Van Allen, of Johns Hopkins, held the chair of the UARRP. The NACA Special Subcommittee was chaired by Harry Wexler of the U.S. Weather Bureau, and included as members Van Allen, future NASA Associate Administrator Homer Newell then of the Naval Research Laboratory, and Joseph Kaplan, who later chaired the U.S. International Geophysical Year Committee. Several other people served both groups, "In fact, many meetings were held consecutively with practically the only changes being the presiding officer and the secretary."[54]

In early 1953, Langley established a study group to consider the details of a hypersonics program. The three-man group included O'Sullivan, and their report recommended that a hypersonic research aircraft be built, supported by rocket model tests from Wallops with the test vehicles to be recovered from the Sahara Desert.[55] Given his connection with Langley's management, his seat on an NACA subcommittee, and his association with military and university scientists outside the NACA, the fact that O'Sullivan (one of those responsible for the early organization of Wallops Station) could facilitate PARD's entry into atmospheric science research came as no surprise.

It is interesting to consider that in April 1958, Smith J. DeFrance, Langley veteran and long-time head of Ames Laboratory, wrote a letter to Robert Gilruth stating that, "the staff of Ames Laboratory is anxious to take advantage of the powerful research technique afforded by the rocket flight-test facilities at the Wallops Island field station," and asking for basic information about the base and PARD's operations.[56] In November, Ames engineers paid a visit to the station.[57] Ames had little contact with Wallops during their early existence. Though originally staffed by Langley trained personnel, Ames' situation on the West Coast made utilization of the Virginia range impractical. Also, the growth of an institutional culture at the California lab created occasional frictions between Langley and Ames, though generally the relationship was one of "friendly rivalry."[58] It thus seems likely that, through O'Sullivan and others, the planners of the International Geophysical Year knew more about Wallops' and PARD's capabilities at an earlier date than did some researchers within the NACA itself.

During the first years of the upper atmosphere research effort the scientists used the converted V-2s, the Naval Research Laboratory's Viking and Aerobee rockets, and the Rockoon system to conduct their experiments.[59] As the three former operated with liquid fuel and the latter proved too inaccurate to launch from land, Wallops contributed little to this phase of the research. By 1958, though, budget constraints forced the consideration of a less expensive booster. L.M. Jones, of the University of Michigan, consulted with O'Sullivan who pushed the Nike-Deacon combination already in use at Wallops. Jones wanted a system that could reach 250,000 feet with a 50 pound payload. O'Sullivan had previously done the necessary calculations and promised Jones 400,000 feet.[60]

On 8 April 1958, the first atmospheric sounding rocket launched from Wallops flew for the University of Michigan. Though the project was funded by the Air Force Cambridge Research Center, this military interest did not hinder project access to the base.[61] The I.G.Y. Committee took quick notice, especially since the use of the Nike-Deacon reduced by a factor of 10 the cost of an experiment previously conducted with an Aerobee. Two successful test flights thus put Wallops into the I.G.Y. program and on the road to an entirely new mission.[62]

The U.S. Weather Bureau also quickly capitalized on this new research capability. Assisted by the Office of Naval Research, the Bureau was looking for a new hurricane detection method. The accidental discovery of such a storm during a rocket flight from White Sands inspired the idea, and the economy of PARD's Nike-Deacon attracted the attention of Bureau Chief Francis W. Reichelderfer, also a member of the NACA Main Committee.[63] He arranged a 20 October meeting with O'Sullivan, and the NRL's John Townsend and Leslie Meredith. The result of this meeting was Project Hugo, a plan to launch Nike-Cajun rockets with a camera package as payload. After taking a series of pictures, the camera package would descend by parachute for recovery by the Navy. The film could then be examined for hurricanes, providing advance notice of their approach. The system sounded good, but for a number of reasons, proved unreliable. Despite an effort watched over by Robert Krieger himself, the first and only successful test of Project Hugo did not occur until 5 December 1958. The pending development of orbital weather satellites promised an easier way to do the job, a vehicle out of Wallops' field, but the Weather Bureau would return to the Station later.[64]

Not all of PARD's early forays into cooperative scientific research went smoothly. In mid–1955, Dr. S. Fred Singer of the University of Maryland proposed a series of research flights sponsored by the National Security Agency. The design of the new Terrapin rocket caused some friction between Singer and PARD, as did the lack of requisite paperwork between the NSA and the NACA. Several memos buzzed back and forth within Langley, and although the project was satisfactorily concluded, it pointed out the need for a new set of procedures at Wallops to facilitate the nascent scientific program.[65]

Most of Wallops efforts in the pre–Sputnik era came at the behest of American organizations. U.S. military, industrial, or collegiate customers monopolized the resources of the PARD. Few foreign projects came to the base. Two projects of interest to the North Atlantic Treaty Organizations's Advisory Group for Aeronautical Research (AGARD) were conducted in 1951 and 1954. Joseph Shortal, who replaced Gilruth as PARD Chief in 1951 reported that he, "presented a paper on the rocket model and Helium Gun [testing] techniques at the Fourth General Assembly of AGARD meeting in the Netherlands in May 1954."[66] Representatives of AGARD did not receive an invitation to visit Wallops until 1959.[67]

It appears that the only non-American group to use the Station during the NACA era was a team from the Canadian Armament Research and Development Establishment Test Range. These engineers encountered difficulty tracking rocket model tests of the CF–105 fighter aircraft. They received permission to launch two CF-105 models at Wallops and came away "impressed" by the Wallops radar operators' ability to quickly acquire and track the model.[68] With all of the military tests underway at the base during this period, especially ICBM and other nuclear related research, cooperative work with international organizations did not rank highly on PARD's priority list.

The increasing speeds and distances associated with the hypersonic program pushed capabilities of the tracking and data acquisition equipment to their limit. The first several years of the program resulted in such an increase in booster performance that, "minimum improvements and a loss of accuracy had to be accepted..."[69] The problem not only concerned increasing the range and sensitivity of the radars involved in tracking the test flights, but also focused on the sensors aboard the models that generated the data, and the telemetry systems that relayed the data to the engineers on the ground. Wallops received most of their radar equipment either from the military as surplus, or directly from the military's suppliers. As radar systems constantly changed to meet increased levels of military needs, those increased abilities found their way into Wallops' equipment, requiring minor IRD modifications. The telemetry systems retained the complexity that had become a trademark of the IRD's specialized work, however, and this equipment continued development on a largely in-house basis. The unique requirements of the hypersonic program called for devices not easily found from industry. The program sparked "an instrument development program for high-speed and high-altitude measurements that was to continue far into the space age."[70]

That age commenced sooner than expected.

NOTES

1. Due to the fact that each research flight requires the launch of several small rockets to calibrate tracking systems, the absolute total number of launches from Wallops is impossible to state exactly. The number stated here is from Berl Brechner, "Space Island," *Air & Space Smithsonian*, April/May 1989, 62. The figure is corroborated by calculations from several sources including: Jane Van Nimmen, Leonard C. Bruno, and Robert L. Rosholt, *NASA Historical Data Book*, vol. 1 (Washington, D.C.: National Aeronautics and Space Administration, 1988), 480 (hereafter cited as *Data Book I*); Table by Wallops Public Affairs Office, June 1979, in file box "J. S. Palmer's Old Records – Historical," in the Wallops Flight Facility Records Collection, Wallops Island (collection hereafter cited as WFFRC). This file box contains assorted typed and handwritten tables depicting yearly launch records, most of which agree generally if not exactly.

2. John Logsdon, "Opportunities for Policy Historians: The Evolution of the U. S. Civilian Space Program," in *A Spacefaring People: Perspectives on Early Spaceflight*, ed. Alex Roland (Washington, D.C.: NASA, 1985), 81.

3. The first important period, 1944–1951, involved the establishment of the base and its early development. The third important period, 1979–1983, centered on the merger of Wallops with Goddard Space Flight Center, and the retirement of many NACA veterans.

4. Alex Roland, *Model Research: The National Advisory Committee for Aeronautics, 1915–1958*, vol 2 (Washington, D.C.: NASA, 1985), 394–95. Unless otherwise noted, citations for Roland hereafter will refer to either volume one or two of this work.

5. Roland, I: 5–29. See also, James R. Hansen, *Engineer In Charge: A History of Langley Aeronautical Laboratory, 1917–1958* (Washington, D.C.: NASA, 1987), 1–5.

6. Hansen, 9–16; quote is on page 11. See also, Roland, I: 79–81.

7. Hansen, 58–62. Hansen also notes that, "forty–six Langley workers died of influenza between September 1918 and January 1919," (page 18).

8. Hansen, xxxii-xxxiii.

9. Ibid., 325. See also, Howard E. McCurdy, *Inside NASA: High Technology and Organizational Change in the U.S. Space Program*, (Baltimore: Johns Hopkins University Press, 1993), 25–34.

10. Roland, I: 147–66.

11. Edwin P. Hartman, *Adventures In Research: A History of Ames Research Center, 1940–1965* (Washington, D.C.: NASA, 1970), 32. See also, Roland, I: 249–50; McCurdy, 25–60.

12. Hansen, 159.

13. Roland, I: 196.

14. Ibid., 108–23.

15. "The transonic region refers to that area between mach .7 and mach 1.3 where a plane encounters mixed subsonic and supersonic airflow." Richard P. Hallion, *On The Frontier: Flight Research at Dryden, 1946–1981* (Washington, D.C.: NASA, 1984), 4. See also, Roland, I: 199; and Hansen, 220, 249–56. Hansen gives a good account of the forces encountered at Mach 1.

16. Joseph A. Shortal, *A New Dimension: Wallops Island Flight Test Range: The First Fifteen Years* (Washington, D.C.: NASA, 1978), 4-6. See also, Hansen, 257-58. The fourth method used to obtain transonic data did not find wide spread use. The wing-flow method involved mounting a small model atop the wing of an aircraft which would then fly close to Mach 1. The airstream close to the wing would go supersonic allowing the research model to experience what the aircraft carrying it could not.

17. Shortal, 9; Roland, I: 252-53.

18. Ibid., 23-4.

19. Ibid.

20. Ibid., 25-32.

21. Shortal, 29.

22. Ibid., 48. See also, Doug Garner, "Seeking Guidance," in *Air & Space Smithsonian*, October/November 1993, p. 80-83; for an account of early operations at Wallops.

23. Shortal, 60-66, 98, 107, 118-22.

24. Ibid., 66-69.

25. "Victory reveals existence of secret missile base," *New York Times*, 24 September 1946. Shortal, 146. See also "Spinak, et al.," Oral History Interview, Tape la: 140.

26. Shortal, 73, 104-5, 115-16.

27. Hansen, 269-70; Shortal, 93-7.

28. Biographical file #001246, "Robert L. Krieger," in the Biographical Collection of the NASA History Office, Washington, D.C. (This repository hereafter cited as NHO). See also, Shortal, 94-95.

29. Shortal, 93-95; on page 68 there is a photograph from October 1945 showing Ira Abbott and John Stack, among others, at Wallops "for a flight operation."

30. Roland, I: 197-98.

31. Ibid., I: 498; Shortal, 104-6.

32. Shortal, 115.

33. Ibid., 141-43, for the helium gun, an apparatus that launched small models with a blast of pressurized helium. On page 183 Shortal relates that 20 people were assigned to the Research Section "which handled all flight operations and the Preflight Jet;" 45 were assigned to the Mechanical Services Unit which maintained the equipment and the base; and 10 were assigned to the Administrative Unit to handle paperwork.

34. Ibid., 186.

35. Ibid., 237-40; Hansen, 269-70.

36. Shortal, 160, 202.

37. Hansen, 348. On page 344 Hansen notes, "Generally speaking, aerodynamicists considered speeds above Mach 5 as hypersonic, since this was the supersonic speed at which aerodynamic heating seemed to become vitally important in aircraft design."

38. Ibid., 356, 362; Roland, I: 200-06. Roland notes on page 253 that the NACA's lack of a clear mandate to pursue rocket research left the investigation of this device largely in the hands of the military.

39. Shortal, 237, remark is attributed to Harry J. Goett, who later became the first director of the Goddard Space Flight Center.

40. Ibid., 241.

41. Ibid., 237-41. Note that unlike the earlier X-series aircraft, no models of the X-15 appear to have been tested at Wallops.

42. Ibid., 116-18, 293-4.

43. Ibid., 170, as to Naval opposition to Wallops increased range, "Langley was informed privately that no relief could be expected as long as the incumbent CINCLANT was in command." See also pages 293-4; and for conflicts with civil air routes, page 116.

44. Ibid., 297-8. "Spinak, et al.," OHI, Tape 1a: 140. For a more detailed description of the housing problem early on at Wallops see: Memorandum, R. R. Gilruth for E. H. Chamberlin, 28 February 1946; Memorandum with enclosures, W. Calvert Roberts to H. J. E. Reid, 9 May 1946; Memorandum, ___ W. White for files, 23 May 1946; Memorandum, H. J. E. Reid to NACA Headquarters, 17 May 1946; Memorandum, H. J. E. Reid for R. E. Ulmer, 9 September 1957; Memorandum, Kurt Berlin for the record, 4 March 1959. All are in folder "Wallops, January - June 1946 [sic]," Record Group A181-1 "Correspondence Files, Wallops Island," in the Langley Research Center Historical Archives. Date on the file should read 1965. This collection hereafter cited as RGA181-1(C).

45. Shortal, 302; Hansen, 270; Roland, I: 264-65. A very revealing insight into NACA - Congressional relations can be seen in: U.S., Congress, House, Committee on Appropriations, *Independent Offices Appropriations for 1957, Hearings before a Subcommittee of the House Committee on Appropriations*, 84th Cong., 2nd sess., 5602-03H, p. 984. Subcommittee Chairman Albert Thomas accused NACA Director Hugh L. Dryden of having "people hanging out of windows and on the roof." Unless otherwise noted, all Congressional records cited hereafter are in the "Congressional Records Collection" in NHO.

46. Shortal, 302; Hansen, 270.

47. Shortal, 427.

48. Ibid., 83, 441.

49. Ibid., 442, 446. For additional information on the needs of the hypersonic program see: U.S., Congress, House, Committee on Armed Services, *Hearings before Subcommittee #3 of the Committee on Armed Services on H.R. 2581 and H.R. 2123*, 84th Cong., 1st sess., 5502-07H, p.360; U.S., Congress, House, Committee on Appropriations, *Hearings before a subcommittee of the Committee on Appropriations on the 2nd Supplemental Appropriations Bill for 1955*, 84th Cong., 1st sess., 5503-03H, p. 325. Memorandum, Joseph E. Robbins to Langley, 30 April 1957, "1959 Budget Estimates," in folder "September 1956 - April 1957," in Record Group A181-1 "Special Files - Wallops Island," in Langley Research Center Historical Archives. This collection hereafter cited as RGA181-1(S).

50. Shortal, 447-50.

51. Ibid., 387-90.

52. Rip Bulkeley, *The Sputniks Crisis and Early United States Space Policy* (Indianapolis: Indiana University Press, 1991), 48. Constance McLaughlin Green and Milton Lomask, *Vanguard: A History* (Washington, D.C.: Smithsonian Institution Press, 1971), 6.

53. Shortal, 251. As NACA committee were populated by people representing many different institutions, the need to assign a NACA employee to represent the organization on its own subordinate committees was common.

54. Bulkeley, 48, 92; Shortal, 252.

55. Hansen, 351-53.

56. Letter, S. J. DeFrance to R. R. Gilruth, 25 April 1958, In folder "Wallops, March - December 58," RGA181-l(C). This file also contains a reply to the letter from H. J. E. Reid. (Shortal notes this letter in a different context on page 669.) For DeFrance's background see: Hartman, 26.

57. Memorandum, Charles B. Rumsey for Floyd L. Thompson, 25 November 1958, in folder "September - December 1958," in RGA181-l(S).

58. Hallion, 14. It should be noted that Ames eventually developed a closer relationship with the High Speed Flight Station at Muroc Dry Lake, though this took time. Unlike the relation between Langley and Wallops, the one a part of the other in a subordinate position, Ames and HSFS were administratively separate until 1981. Indeed Ames lost its flight research program to HSFS in 1959. See also: Hartman, 315.

59. The Rockoon consisted of a Deacon rocket with payload suspended beneath a balloon. Once the balloon reached a given altitude the Deacon ignited and launched. Unfortunately, there was no way to tell in what lateral direction the rocket would fly, so it could not be launched anywhere near a populated area. Shortal, 252, 401.

60. Shortal, 401-3.

61. Ibid., 400, "Funding for most of the upper atmospheric research was provided by the Armed Forces."

62. Ibid., 403.

63. Roland, II: 433. Recall that the Bureau's Harry Wexler chaired the SSUA.

64. Shortal, 459-64. Memorandum, William J. O'Sullivan to Associate Director Headquarters, 12 March 1956, in folder "January - December 1956," in box #4 "Reference Material for the Book Entitled *A New Dimension*," in WFFRC. This box hereafter cited as Wallops box #4. The Nike-Cajun was an uprated version of the Nike-Deacon.

65. Shortal, 504-7. The Terrapin rocket was a new configuration of the Deacon and a T-55 motor. Memorandum, William J. O'Sullivan for Associate Administrator, 10 January 1956; Memorandum, H. J. E. Reid for NACA Headquarters, 13 January 1956; Memorandum, H. J. E. Reid for NACA Headquarters, 5 March 1956, all in folder "January - December 1956," in Wallops box #4. For the NACA desire for regimented paperwork, a by-product of Executive Secretary John Victory's style, see Hansen, 24, 28-33.

66. Shortal, 220, 369.

67. Ibid., 617.

68. Ibid., 457-58.
69. Ibid., 292.
70. Ibid., 292, 302.

Chapter 2

SPUTNIK, NASA, AND INDEPENDENCE

The dawn of the Space Age, and the start of the Space Race, occurred in the darkness of the Russian night on 4 October 1957. The successful launch of *Sputnik 1* opened a new front in the Cold War and turned the idea of space operations from science fiction into science fact. The expressions of "shock" by American politicians, scientists, engineers, and the public at large seem almost cliche through repetition. However shocking to the public and politicians, the Soviet achievement should not have been too big a shock to knowledgeable professionals. The Soviets announced their intent to orbit a scientific satellite as part of the International Geophysical Year more than two years before the fact.[1] American researchers realized that "going orbital" not only involved a relatively simple extension of emerging technology, but also that someone would do it soon. ABMA, NRL, RAND, various groups within the NACA (including PARD), and others, all nursed orbital visions of varying priority.[2] Of course, they generally assumed that the first beeps from space would be generated by an American transmitter, but after the demonstration of Soviet technical capability represented by Russia's nuclear program they largely took the success of Sputnik in stride.[3]

President Eisenhower also evinced little concern about the Soviet accomplishment. He placed a greater value on the program to develop an operational ICBM and, supported by a slowly growing number of military officers, determined that nothing should stand in the way of obtaining this new weapon. The importance of speedily executing this program was emphasized by the launch of *Sputnik 2* on 5 November. Weighing more than half a ton and carrying a dog as one-way supercargo, this new satellite exhibited a launch capability beyond expectations.[4] The possible substitution of a nuclear warhead for Laika the dog generated concern nationwide as Americans realized that the wide oceans no longer provided security from sudden attack. In a televised address two days later, Eisenhower attempted to calm nerves by calling attention to the strength of American forces, citing progress in the ICBM program, and appointing Dr. James Killian, Special Assistant for Science and Technology. Later that month, Eisenhower put Killian in charge of the President's Science Advisory Committee (PSAC), and appointed William Holaday Director of Guided Missiles.[5] Generally, though, Eisenhower down played the significance of the Russian satellites, and recommended only moderate funding increases for missile research.[6]

Political adversaries were quick to utilize the Sputniks to criticize the President and further their own agendas. Senate Majority Leader Lyndon

Johnson emerged as the most visible figure on Capitol Hill during the post-Sputnik scramble. Looking toward the 1960 presidential campaign, Johnson saw the opportunity to push for a boost in defense spending while playing to a national audience. On the day after *Sputnik 1* went into orbit, he began organizing an "Inquiry into Satellite and Missile Programs," by the Preparedness Subcommittee of the Senate Armed Services Committee.[7] During the Subcommittee's hearings Johnson examined a wide range of defense and space related issues, cultivating the idea that Sputnik represented a dangerous challenge to American security, and attacking Eisenhower's conservative fiscal policies. Though Johnson took care to conduct the hearings in a bipartisan manner, they clearly portrayed the Administration in an unfavorable light.[8]

Despite Eisenhower's best efforts and personal feelings, public concern with the space and missile issue grew. On 6 December 1957, in the full glare of the media spotlight, a Vanguard rocket exploded just after liftoff during the first U.S. attempt to orbit a satellite. This failure, despite the fact that the Vanguard system was still undergoing tests, combined with the previous Russian successes to confirm the nation's worst fears. The Soviets, naturally, made the most of the propaganda opportunity, and Eisenhower began to realize that the political situation could not be safely ignored.[9] Several groups within the Administration set to work.

Widely blamed for letting interservice rivalries permit the Russians to obtain their technological lead, the Defense Department took steps to correct the problem by creating the Advanced Research Projects Agency (ARPA). This agency began the difficult task of coordinating the various military space programs.[10] The Air Force portrayed space as just an extension of its operational arena. The ongoing program of heavy booster development for ICBMs and its involvement with the X-series aircraft gave the airmen powerful arguments for the assumption of all U.S. space endeavors. The Army maintained their argument that missiles represented a form of long-range artillery, and used the Vanguard accident to showcase their space abilities. On 31 January 1958 *Explorer 1* ascended into orbit atop a Jupiter-C booster designed by Wernher von Braun's rocket team. This group of transplanted German scientists, a large part of the group that designed the V-2, had been working for the U.S. Army since the end of the war and represented one of the most technically advanced cadre of rocket engineers in the country. The Navy, while interested in the potential usefulness of applications satellites for its far-flung operations and stung by the Vanguard failure, seemed more interested in its Polaris program and in not allowing another service to monopolize space. Indeed, this mutual jealousy characterized all three services; none wished to be shut out of the role of space defender, a role certain to entail an increase in funding.[11] All jockeyed for position. Both President Eisenhower and Senator Johnson, however, had other ideas.

24

Eisenhower, already wary of the growing "military-industrial complex," intended that there be separate military and civilian space programs. Johnson, with an eye toward political ramifications (both at home and abroad) concurred. While riding herd on various Congressional proposals, Johnson waited for Eisenhower to make the first move.[12] The task of crafting the Administration's civilian space program rested with Killian and PSAC.

The upper echelons of the NACA, long opposed to any "Buck Rogers" projects, initially felt no more concern over Sputnik than did Eisenhower. The Executive Committee had met at the Wallops Base on 19 September 1957 and obtained the latest information on the hypersonic research program.[13] The Main Committee held its annual meeting on 10 October, one week after the Soviet launch, and did not discuss the matter.[14] This lack of high level concern stemmed from a combination of factors.

For years the notion of space operations, piloted or not, received the label of science fiction from the public, politicians, and many professionals alike. An in-depth, publicly financed, space research program would not have been approved by the fiscally conservative Eisenhower or a Congress that reflected the opinions of its largely unimaginative pre-Sputnik constituency. PARD and the other interested groups within the NACA fought to justify their limited forays into astronautics to their own Headquarters, who in turn never forgot that the NACA's budget underwent severe scrutiny before the often critical Albert Thomas and his House Subcommittee on Independent Offices Appropriations.

The primary reason the hypersonic program received funding stemmed not from its scientific and engineering potential but from its obvious military significance. The military monopoly on missile research, and the attendant political maneuvers between the services, served to keep the NACA's official goals within the atmosphere, a situation not necessarily to the disliking of ranking Committee members. Technically "innovative" but by nature methodical and by necessity politically conservative, most NACA decision makers regarded space as primarily a military area, and likely to remain so for the foreseeable future. If research eventually led the NACA into space the subject could be dealt with at that time; meanwhile why irritate the military, a prime customer and powerful political ally?[15] Not all within the organization viewed the situation in this light however.

Many throughout the NACA disliked the clean-up research necessitated by their close relationship with the military. Some saw space research as a way to return NACA to its roots by emphasizing basic engineering research. There also existed the increasing perception that aeronautical research "was reaching a point of diminishing returns," and that if the NACA could not expand into astronautical research Congress might just decide that expanding military and industrial research capabilities made the NACA superfluous.[16] Years of declining budgets and escalating criticism made many nervous.

Several groups within the NACA labs quietly crafted plans for space research. Abe Silverstein and his associates at Lewis Lab began experimenting with liquid hydrogen and other potential chemical rocket fuels, and unobtrusively studying electric and nuclear propulsion[17]. Ames and Langley studied lifting bodies and hypersonic successors to the X-15 that incorporated space flight capabilities.[18] PARD also looked to space with some of its engineers already having done "back of the envelope calculations" pertaining to obtaining orbital velocities with their rockets.[19] Robert Gilruth later wrote, "I can recall watching the sunlight reflecting off the Sputnik 1 carrier rocket as it passed over my home on the Chesapeake Bay in Virginia. It put a new sense of value and urgency on the things we had been doing."[20]

The 18 November meeting of the NACA Committee on Aerodynamics (held aboard the aircraft carrier *U.S.S. Forrestal*) paid much more attention to Sputnik than had the Main Committee the previous month. "The big question to be answered now is how can these views [on accelerating space research] be put across to the NACA and to the Government in order that the NACA be recognized as the national research agency in this field, and be provided with the necessary funds. ... the NACA should act now to avoid being ruled out of the field of space flight research." The Committee suggested spotlighting the hypersonics program in general and the X-15 project specifically in order to make the case.[21]

This committee, at least, recognized the need for alacrity. Eight days after this meeting, NACA Chairman James Doolittle testified before Johnson's Preparedness Subcommittee. Interestingly, Doolittle referred to Wallops as "a missile-testing laboratory," during his testimony, in contrast to a 1951 NACA press release that emphatically stated that "this is an aerodynamics range, not a proving ground for missiles."[22] Throughout this period, in testimony before many committees and in public statements, Wallops was rarely referred to directly. The programs in progress there received much attention, but the potential offered by the facility and its staff seldom appeared in print. This should not be taken as a sign that Washington knew nothing of Wallops. For example, on 19 November, Acting Engineer-in-Charge John Palmer received a phone call "at quitting time," from the Executive Officer at Chincoteague Naval Air Station. He advised Palmer that a committee from Washington had conducted an inspection of the Navy base that day and wished to inspect the NACA facility next. Not familiar with the "Special Committee on Range Facilities," Palmer called Krieger. After failing to contact Gilruth, Krieger contacted Buckley and the two travelled to Wallops and met the committee on 21 November. The visitors turned out to be a high-level group from the Office of the Secretary of Defense studying the "long-range, over-all situation in regard to adequacy of test ranges in order to assure that facilities were available when needed and to prevent duplication, conflicts, etc." The group's mission also included scouting a location for a new test range as, "the services are being forced out of the

Delaware – New Jersey coastal area because of the density of population." This was one of several groups that visited the Station during this time.[23]

The Main Committee established a Special Committee on Space Technology on 21 November chaired by H. Guyford Stever from MIT, included such luminaries as von Braun and Van Allen, and also placed Gilruth on the roster. This committee served to coordinate and champion the NACA's attempt to expand into the new arena. The perception began to grow within the organization that space research might be an all or nothing proposition. If they could not win the civilian space mission, they might be absorbed by the group that did; one way or the other, changes loomed on the horizon.[24] A number of studies appeared promoting the NACA and setting forth its qualifications and requirements for assuming the space program. One such study noted that, "the Pilotless Aircraft Research Station, ..., is now being used almost exclusively on hypersonic and space flight problems.[25] In testimony before Congress witnesses estimated PARD activities to be "90% Space Research."[26]

The NACA won round one by convincing Killian's Advisory Committee on Government Organization that it should be assigned the space mission. "We recommend that leadership of the civil space effort be lodged in a strengthened and redesignated National Advisory Committee for Aeronautics."[27] The NACA's history of close relations with the military and the applicability of its programs and facilities to space research led Killian's Committee to recommend NACA over other contenders such as the Atomic Energy Commission (AEC), a proposed "Department of Science and Technology," or private contractual arrangements. On 5 March 1958 Eisenhower gave the recommendation his approval. The NACA prepared for round two of the contest: winning Congressional approval.[28]

In early February Senator Johnson oversaw the creation of the Senate Special Committee on Space and Astronautics, of which he became chairman, to provide an organizational vehicle for Senate input into the issue. The House established a similar committee, chaired by Majority Leader John W. McCormack, the following month.[29] When the administration's bill proposing the expansion of the NACA into a National Aeronautics and Space Agency went to the Hill, these committees conducted the requisite hearings. After addressing concerns about the NASA's relation to military space programs and patent rights, among other issues, the bill passed both Houses and was signed into law on 29 July 1958.[30] The NACA had won the space assignment, but not quite in the form it had desired.

Against their objections the Space Act replaced the old committee system with an administrative system subject to tighter executive branch control and legislative branch oversight. NACA Research Director Hugh Dryden, everyone's expected nominee for the post of Administrator, failed to impress Congress during the hearings on the Space Act and was passed over in favor of T. Keith Glennan, president of Case Institute of Technology and former member of the AEC. The National Aeronautics and Space Administration comprised more than just the NACA. The Vanguard division of the NRL soon transferred in, and NASA assumed the Army's contract with the Jet Propulsion Laboratory in California. After a bureaucratic struggle, the von Braun team came over from the ABMA. The addition of these organizations, each with their own backgrounds and institutional cultures, and the assignment of an outsider to the top post (though Dryden accepted the #2 position), virtually guaranteed that NASA would not be simply a renamed NACA.[31]

The early days of any organization usually involve a fair amount of chaos, and NASA's proved no exception. Integrating established components into a new structure while in the public spotlight and under the pressures of a perceived, if undeclared, space race with national survival seemingly at stake, promised to make Glennan's task a difficult one. His early impression of the NACA seemed consistent with that of many outside the agency. "Although NACA had on its roster some very fine technical people, it had been an agency protected from the usual in-fighting found on the Washington scene." Management he described as, "reasonably able," but they, "had relatively little experience in the management of large affairs."[32] An Ad Hoc Committee on NASA Organization, chaired by Ira Abbott, had been instituted in April 1958 to formulate the NACA's vision of the new space agency. Glennan reviewed the Committee's report with Dryden and the top NACA leadership, then let a contract to a management consulting firm, McKinsey & Co., to review and expand upon it from a perspective outside the agency.[33] The McKinsey Report took fire as having "rubber-stamped" the Abbott Committee report, a critique not altogether unwarranted. One of the many similarities turned out to be the role of Wallops in the new organization.[34]

In order to minimize interference with ongoing aerodynamic research (especially the militarily vital heat transfer and hypersonics projects) by the new space agency, the organizational plans called for placement of Langley, Ames, Lewis, and HSFS, under one branch of NASA specializing in aeronautical research. A new space research center, staffed by the Vanguard group and a substantial portion of NACA's "space enthusiasts," including Gilruth and many PARD veterans, would carry out the civil space program. Toward this end the plans called for the separation of Wallops from Langley and its situation as an appendage of this new center.[35] Indeed, the possibility of locating the new center at Wallops was briefly discussed. However, the lack of sufficient local infrastructure to support the proposed large facility, and the desire to keep the center close to Washington (for political and logistical reasons), doomed this prospect. Nevertheless, a facility "90%" devoted to space research could only be placed within the space portion of the agency.[36]

The perception of Wallops' role within NASA differed from that of either the old aeronautical centers or the new space centers. Wallops and both the

Atlantic and Pacific Missile Ranges constituted "service centers" in the eyes of NASA leaders, bases at which the research centers could conduct experiments, rather than independent centers conducting research of their own. Wallops and the NASA facilities at Cape Canaveral therefore appeared in the organizational charts directly under the space research center's supervision.[37]

NASA located the new space lab on a parcel of land outside Washington obtained from the Department of Agriculture. Operating first as Beltsville, then as Goddard Space Flight Center, the facility existed for several years largely as a paper organization, until the physical plant could be built. The Vanguard group continued to operate out of the Naval Research Laboratory, and Gilruth's Space Task Group remained at Langley. Though nominally a part of Goddard, "You'll find very few people today who'll realize they were working for Goddard back then, because they weren't."[38] Goddard, with buildings under construction and attempting to integrate disparate research programs in one organization, could not begin to direct and support Wallops operations as well. As late as April 1959 Space Center personnel were still in the process of planning their own "future activities at Wallops," and even though many researchers officially at Goddard, especially Gilruth, knew Wallops well, administration of the base could not efficiently be done from Beltsville.[39] NASA could not put its programs on hold to allow Goddard time to mature, and the projects at Wallops required expedition.

The first NASA staff conference, held in April 1959, devoted one session to "A Critical Examination of the Organizational Requirements of NASA." A part of this session examined the "Place of Wallops and NASA Staff at Canaveral in the Organization." NASA recognized that, "both Wallops and Canaveral will be concerned with firing the products not only of Beltsville but of other NASA activities." However, "it has not yet been firmly decided where in the organization Wallops and the NASA activity at Canaveral will report."[40] Perhaps, but the matter surely drew attention. The organizational chart released the following month shows Wallops as an independent entity, coequal with Goddard, under the direct jurisdiction of Abe Silverstein's Office of Space Flight Development at Headquarters.[41] (See appendix 2)

While Wallops thus occupied a new place in the organization, Langley continued to provide administrative, logistical, and engineering support to Wallops for several years. Officially separate, Wallops for all practical purposes continued to operate much as it had all along, as an appendage of Langley.[42] The effect of Sputnik and the NACA to NASA upheaval therefore proved to be a curious mixture of transformation amid business as usual at the Wallops Station. It has been observed that the people who left work 30 September 1958 NACA employees, returned to work on 1 October NASA employees. The launch log at Wallops does not reflect this change though; projects continued as before.[43]

Personnel from Langley and Wallops laid plans for a massive expansion of the facilities both on the island and on the Virginia mainland opposite it. The Station represented a prime national space asset at a time when such facilities seemed scarce. Congress appeared willing to fund almost anything to get the suddenly urgent program on track. PARD took advantage of this opportunity by preparing a wish list that included a causeway providing direct access to the island, launch equipment for Thor and Jupiter class liquid-fueled boosters, and service and administrative buildings requiring the purchase of over 1000 acres of land. The estimated cost of the program ran to $24 million.[44] The House quickly appropriated $1,000,000 for enough rockets to maintain the schedule of operations at Wallops, and more seemed forthcoming.[45]

Reality soon intruded. Neither President Eisenhower or Representative Thomas intended to allow space projects to bust the budget. Despite the fact that, in these early days, "Congress always wanted to give us more money," NACA officials in April 1958 found themselves in a familiar setting: before the Senate Subcommittee on Independent Offices asking for a restoration of funding cuts made by the House.[46] Apparently some in Congress believed an incremental approach preferable to a crash program. Also, the military began to realize that NASA would not be just a re-named NACA, subservient to the armed forces' desires. Military planners originally worried that the creation of the NACA might draw funds away from their programs. NASA ignited those same concerns, only magnified, in a new generation of senior officers, who began to view the space agency as a competitor.[47]

The first thing deleted from PARD's list was the capability to launch large liquid-fueled boosters from the island. The work load at Wallops remained heavy and showed little sign of abating anytime soon. The plan called for a new launch complex consisting of two pads and a centrally located blockhouse from which to control firings. If one of these pads possessed the capability to support Thor or Jupiter rockets the civilian range would gain an immediate access to orbital spaceflight without having to coordinate activities with the military controlled ranges. Purely civilian science projects could be conducted without interference with, or from, sensitive activities at the Cape.

Both the Congress and the military balked at this plan. Duplication of facilities, especially those requiring significant amounts of money to build, remained intolerable to most legislators, and with expensive complexes rising at the Cape and at Vandenberg Air Force Base, Congress saw little need for such a large scale increase in Wallops' capacity.[48] Military witnesses reinforced this argument. In response to a direct question about approval of the Wallops expansion one admiral replied, "If they want to keep it on the small sized rockets scale—fine; but if they want to put satellites into space, it should not be done from Wallops Island."[49] The officers did not want small projects, like those that generally utilized sounding rockets, interfering with the operations of the large ranges, but, they also did not want Wallops to expand into a competitive position.[50]

The Stever Committee, in the course of its evaluation of the NACA's space resources, examined Wallops and considered the role the base might play. They recognized that a launch site situated at a latitude of 37.5 degrees would not be practical for lunar or interplanetary flights. Though supportive of the plan to equip Wallops for boosters "up to the size of the Redstone," the Committee believed the range best suited "for special work on techniques and components in support of the [civilian space] program." The Atlantic and Pacific Missile Ranges (the Cape and Vandenberg) would serve as the nation's primary spaceports.[51] This less than enthusiastic endorsement did not offset Congressional and military concerns, and the expansion underwent "re-evaluation."[52]

Another reason for the re-evaluation came in the person of the Administrator of the new Federal Aviation Administration (FAA), Elwood Quesada. In May 1958 the Air Force Special Weapons Center approached PARD requesting help with the ARGUS program. Briefly, this experimental program sought to determine the radiation characteristics of small atomic explosions in near space; how the radiation generated by an explosion would interact with the Earth's magnetic fields, rates of decay, and questions of that nature. The program required the detonation of a series of bombs at an altitude of 300 miles. Sounding rockets launched from three locations were needed to record the data. The nature of the experiment required a tight launch schedule as environmental readings just prior to, and immediately following the explosions (as well as a third launch after a predetermined period of time) were needed. The five stage rocket developed by PARD for the hypersonics program fit the needs of the Air Force, and Wallops set to work training launch crews and assembling rockets. Originally not intended to serve as a base for launchings during the operation, the fast approaching deadline mandated by an international ban on atomic air bursts (scheduled to go into force in September 1958) forced planners to forgo a second launch site outside the continental U.S., and included Wallops with Canaveral and Puerto Rico in the firing plan.[53]

The program, conducted in August and early September, fulfilled Air Force expectations, and impressed the officers involved with the speed and skill of the Wallops operation. During the course of the launches, however, the military ordered civil air traffic rerouted to ensure safety and security. The usual procedures for obtaining range clearance for Wallops firings included plenty of advance notice to air traffic controllers, notice that this project could not provide. Because of this, "the pilots and the airlines 'raised the roof,'" and incited a backlash not against the Air Force, but against Wallops.[54]

Difficulties concerning clearance for long-range firings had been occurring for some time as Wallops' capability grew, and the imminent expansion of the base intensified the problem. The diversion of flights during the ARGUS support launches brought the matter to a head. Administrator Quesada suggested restricting future long-range operations to the Cape and scaling back the expansion at Wallops. PARD dealt with this problem in a time tested way; they invited their critics to take a tour of the facilities at the base.[55] Civil Aeronautics Administration head James T. Pyle and a group of his associates visited the island on 24 November 1958, observed two launches, and listened to Shortal and Krieger detail NASA's plans for Wallops.[56] On 8 December a NASA group which included Buckley met with Quesada and an FAA contingent which included three CAA members who had taken the tour. "The CAA decidedly was on NASA side during this discussion, ... One of the CAA men stayed after the meeting to tell General Quesada that the use of AMR for the Civilian Space Program would be far worse a problem for the CAA than if the load was divided by the use of Wallops Island." Quesada, described as "versed on missile operation, aerodynamic research," relented saying, "he was not going to oppose NASA programs," but indicated that a workable system of coordinating long-range firings must be developed.[57] Such an agreement was reached in January 1959.[58]

While NASA secured an amicable solution to the problem of range interference with both the FAA and the Navy (concerned about interference with their training areas), the combination of this problem and economic factors led to a decision to drop the large liquid-fueled boosters from Wallops' expansion program.[59] The planned administrative separation of the base from Langley, and the increase in programs on tap still required a large acquisition of land for offices, shops, tracking stations, and housing, even without the Thors and Jupiters.

Even before NASA's debut, procedures commenced to appropriate acreage on the mainland opposite the island. Station personnel made contact with a number of residents to obtain permission to conduct surveys, and with local lawyers to conduct title searches.[60] Wallops' Administrative Officer Joseph Robbins recalled the bad feelings generated at the time: "These people were principally farmers, and they would tell us point blank, 'We don't want you here.'" Not only "unhappy" about losing land, the farmers believed that an expansion of Wallops operations would drive up the cost of labor in the area.[61]

While the Wallops personnel tried to smooth ruffled feathers, several people contacted their Congressional representatives and wrote letters to the local newspapers. A series of polite letters passed back and forth between Administrator Glennan, various other NASA officials, and concerned politicians, including powerful Virginia Senator Harry Byrd.[62] The staff both at Headquarters and at the base understood the situation and sympathized with the local resident's problems, but the NASA program lacked firm definition. No one knew for sure how much land would ultimately be needed for the Station, and though assured that, "we would not be hiring farm help," rumors about the effect of the expansion on the local economy ran wild.[63]

On 9 December 1958 letters went out to twelve individuals notifying them that NASA intended to take possession of plots of land ranging from 11 to 450 acres in size.[64]

At approximately the same time these letters went out, Wallops personnel received the unexpected news that the Navy intended to close the Chincoteague Naval Air Station (CNAS). An outpost of the Norfolk Naval Air Station, CNAS had been expanded by the Navy in November 1942 as part of the early war build-up. In October 1943 the Bureau of Ordnance arrived, and by 1946 added their Naval Air Ordnance Test Station to the base. After World War II the base served mainly as a carrier aircraft training facility.[65]

Spread out over 2000 acres, CNAS was in the midst of renovation when the Sputniks flew. The Navy lengthened one of the three runways from 8,000 to 10,000 feet and constructed several new test facilities, buildings, and a new hangar. A fiscal 1957 appropriation allotted $170,000 for continuing the upgrades.[66] The Bureau of Ordnance requested permanent transfer of the north end of Wallops Island on 5 May 1958 in reciprocation for the naval transfer of an NACA used portion of Moffett Field to the Ames Lab, a request which the NACA denied.[67] Congress refused to allocate a requested $770,000 in fiscal 1958 for CNAS, however, and the Navy decided to economize by closing the base.[68]

The potential effects of the closure on Wallops' operations and the surrounding community concerned all in the area. PARD relied on CNAS to provide emergency medical, air terminal, and weather fax services. The military maintained a restricted airspace zone around Chincoteague which not only covered the NACA operations, but gave PARD an ally when attempts to alter civil air routes occurred.[69] One of the primary factors for choosing the Wallops site had been the presence of CNAS. The loss of jobs and money understandably worried local residents, and served to mute criticism of the expansion of the rocket range.[70]

The advantages offered by acquiring the Navy base did not take long to dawn upon planners at NASA Headquarters. The administrative facilities, shops, and needed acreage, to say nothing of the airfield itself, could be obtained for the proverbial song. After all, the Space Act authorized the transfer of facilities needed for the space program to NASA, and the obvious economy of recycling CNAS, and saving at least some local jobs would undoubtedly impress Congress. Glennan and his staff moved quickly, and found the Navy receptive to the transfer.[71] On 22 January 1959 Glennan formally requested transfer of the base to NASA. At first, the Navy wished to make continued access to the airfield a prerequisite for the transfer. Fearing potential electrical interference with the new, highly sensitive radars planned for Wallops, NASA refused this stipulation, the Navy relented, and the transfer officially took place on 30 June 1959.[72]

Robert Krieger, acting for NASA, accepts Chincoteague Air Station from Navy Captain Toth in change of command ceremonies, June 30, 1959.

The acquisition of CNAS shifted the perspective of many local residents. "The local newspapers were optimistic about a large influx of industry and the possibility of a second Cape Canaveral on the Eastern Shore."[73] Two days after the public announcement of the impending transfer NASA sent letters to most of those expecting to lose land, notifying them that, due to the transfer, NASA no longer needed their land.[74] "We changed [after the announcement of the transfer] from somebody who was no good, to somebody who was real good, because all these people who got laid off all of a sudden were looking for jobs."[75] Unfortunately, many of the high expectations of the community proved premature and too optimistic. NASA already realized that a "second Cape Canaveral" would not be built at Wallops. Agency officials discussed the transfer during their weekly staff meeting on 6 February 1959: "A number of private and public bodies are concerning themselves with the full utilization of Chincoteague in order to minimize the economic impact on the community. ... The consensus was that NASA should firm up the specific restrictions which must be placed on other uses as required by the technical requirements of our operations. To the extent other proposed uses are consistent with these requirements, NASA should be as cooperative as possible."[76]

Some resistance to the transfer arose at Wallops itself, as the amount of land and facilities to be acquired far exceeded the amount required.[77] Everyone involved knew that, for the short-term at least, NASA operations would not rival the Navy's in scope. Accordingly, Headquarters authorized Wallops "to make firm commitments for the hiring of not to exceed twenty-five

(25) of the Navy employees of [CNAS]." The civilian complement of the base numbered around 760[78]. This small number of hires disappointed residents, and generated a formal complaint charging racial discrimination to boot.

The complaint, filed on behalf of an "anonymous individual" by the Worchester County (Maryland) Civil League, came to the attention of both Administrator Glennan and the President's Commission on Government Employment Policy. Maryland Congressman Thomas Johnson also took an interest in the case.[79] An investigation and subsequent report satisfied the Committee, as well as the Civic League and Johnson, for no appeal seems to have been made. Most likely the expansion of Wallops' operations created a sufficient number of jobs, "NASA 1960 plans contemplate an increase of about one hundred in the staff at Chincoteague," to ameliorate the situation.[80] At the least, no other such charge against Wallops appears in the Administrator's Monthly Progress Report during the duration of that document's publication. A number of such charges regarding other NASA facilities do appear, and the Headquarter's Office of the General Council developed an employment non-discrimination policy program in 1960. On 28 September 1960 the Administrator, "submitted a report to the Secretary of the Cabinet and wrote a personal letter to the head of each NASA field installation commenting on NASA's program and urging continued effort in this area." [81]

The acquisition of CNAS served to bring about the return of Robert Krieger to offices at the Station. The imminent separation of Langley and Wallops (after the loss of many of its "space enthusiasts" to the Space Task Group, Langley reorganized the remnants of PARD into the Applied Materials and Physics Division in December 1959), and the improving local conditions brought Krieger back to the Eastern Shore. He moved into the commanding officer's quarters of Chincoteague immediately following the transfer of the base, and set to work reorganizing his charge.[82]

The Mechanical Services Unit, which formerly reported to the Mechanical Services Division at Langley, became the Technical Services Division with William Grant as Chief. The Administrative Unit, which likewise previously reported to Langley's Administrative Officer, also became an autonomous Division under Joseph Robbins. The Research Section, prior to the separation, the only unit under Krieger's direct control, became the Flight Test Division with John Palmer serving as Chief.[83] While these new divisions continued to rely on Langley for an uninterrupted flow of support, they gradually became capable of functioning on their own. "Langley didn't just shed us, they supported us completely, ..., they didn't divorce us until we picked up our own capability, ..."[84] This continuity of support proved crucial to Wallops. Events moved quickly and many important projects accelerated their pace during this period.

The expansion project at Wallops came under the general heading of "Project 2080." This project, one of the earliest commenced by NASA, covered almost all aspects of the physical buildup of the Station.[85] The inheritance of CNAS allowed the reprogramming of funds from land acquisition, construction of mainland facilities, and architect's services, to road improvements around the Station, and completion and modification of buildings at the naval base.[86] This reprogramming resulted in an overall savings of $2.25 million. Scrapping plans for the large boosters saved approximately $1 million, and "lower estimates of instrumentation costs," spared another million. The cost of the expansion project finally settled at $21 million.[87] Wallops' portion of NASA's construction and equipment budget for fiscal 1959 far exceeded that of any other field center, a situation that never again occurred.[88]

Two urgent parts of project 2080 began almost before the project was approved. A causeway allowing paved access to the island remained an unfulfilled dream from the establishment of the base in 1945. Wallops planners became determined that come what may, the causeway would be built, and the ferries (one of which caught fire in 1953, injuring 14 people) and seaplanes (one of which crashed in 1954, slightly injuring the pilot, and engineer Marvin McGoogan) would be retired.[89] The other task concerned the extension of the seawall protecting the equipment on the island. Only a few feet above sea level, storms easily damaged facilities on the island. The NACA fought an on-going battle against the sea and held their own. The expansion of launch facilities and damage to the existing seawall required NASA to move fast to protect their investment.[90] Despite the funding increase to implement Project 2080, local contractors did not experience quite the boom they had anticipated. The scale of many of the tasks involved in building up the Station simply proved too large for local businesses to handle. While smaller contracts often did go to local companies, the big construction contracts by necessity went to large engineering firms not located on the Eastern Shore.[91] Also the window of opportunity for funding big contracts at Wallops turned out to be brief. NASA requested no funds for construction at the base in the fiscal 1960 budget. Hugh Dryden testified that, "At Chincoteague there are, ..., buildings way beyond anything we can foresee for the use of NASA."[92]

Expansion at the base involved more than just expansion of the physical plant. Plans also called for modernized, more capable equipment. The important tracking and data relay equipment figured prominently in the extension of Wallops' space role. Indeed, the sensitivity of the new equipment called for carefully controlling growth in the immediate vicinity in order to ensure optimum performance.[93] Three radars incorporating dishes 60 feet in diameter arose on the mainland opposite the island. Designed and built by MIT's Lincoln Laboratory, two of the three belonged to MIT and operated with funding from ARPA, while NASA operated the third. The NASA dish, designated "Spandar," provided an increased tracking capability required by the continuing hypersonics program and the new space projects.[94]

Telemetry effectiveness improved with the addition of FM/FM, high-gain, and digital systems which vastly increased the amount of data recovered from each flight, and also streamlined the data reduction process.[95] New range control equipment included a refined launch timing system, an enhanced safety command destruct system, and better optical tracking equipment (telescopes and cameras).[96]

Despite the cancellation of the large boosters from Wallops' plans, construction of the new launch complex proceeded. The pad intended for Thor and Jupiter use retained its importance, albeit in a different way than originally intended. The need to meet the accelerated launch schedule provided sufficient justification for building the pad. The second pad in the new complex, intended for a different vehicle, also became an integral component in the program.

PARD researchers working on the hypersonics program successfully utilized more powerful booster designs. Their success with the five stage, Mach 15 vehicle led to a Mach 18 vehicle and studies on ways to expand capabilities further.[97] As a matter of routine PARD engineers kept abreast of new solid motor designs under development by different manufacturers. Unlike many of the payloads, Langley personnel could not produce the solid motors they depended upon. They installed electrical systems, aerodynamic structures, and coupled the motors together in a multitude of combinations, but they lacked the facilities to cast the propellant. They obtained motors, usually with the help of the military, from commercial producers. When one of these companies produced a new or uprated design, PARD checked its characteristics for applicability to the flight research program. In late 1957, after analyzing several modified motors (including the X-248, slated for the Vanguard launcher), the engineers realized that a four stage combination of these existing motors could give them an orbital capability. The payload would be small, true, but in these early days the ability to put anything into orbit meant advancing knowledge. The fact that some questioned the ability of any solid-fueled vehicle to reach orbit also provided a challenge.[98]

Though inexpensive compared to the large liquid-fueled systems, the new solid design represented a substantial increase in cost over the usual boosters in use at Wallops. The public competition between the Army (von Braun, Jupiter) and the Navy (NRL, Vanguard) teams working to launch a satellite made the PARD design unwelcome.[99] It stayed on Langley's drawing boards until mid-1958 when the Air Force showed interest in a new sounding rocket to exceed the performance of the existing Javelin booster.[100] PARD proposed the new solid booster, which could perform as either a sounding rocket or an orbital launcher, and the Air Force accepted. Although not all in NASA greeted it with enthusiasm, the "Solid Controlled Orbital Utility Test System," or Scout, became a long-term member of the agency's stable of boosters.[101]

The Air Force eventually decided to launch their version of the "poor man's rocket" (known as the Blue Scout) from Canaveral, but field level interest in the booster's potential, especially among the space science planners at Goddard, solidified the program at Langley and Wallops. Work on the island's Launch Area #3, the control center, and the upgraded tracking systems necessary to control a launch to orbit, culminated with the 1 July 1960 launch of the first Scout test vehicle. A minor problem gave this first shot a "partially successful" rating, but moving from initial contracts to first launch in about twenty months represented an accomplishment in itself. The Langley engineers soon fixed the problem, and Wallops obtained a significant increase in its operational capability.[102]

As a result of this new capability, a dispute arose within NASA Headquarters over just who should have operational control of the Wallops facility. NASA's original structure included an Office of Aeronautical and Space Research, to which the old NACA centers (except Wallops) reported, and an Office of Space Flight Development to which the space centers reported. The transfer of von Braun's team from the Army to NASA, a decision made by President Eisenhower on 21 October 1959, caused a reorganization of the Space Flight Office. Abe Silverstein, Director of the new Office of Space Flight Programs (OSFP), retained three of his subdivisions while the fourth separated to become the Office of Launch Vehicle Programs (OLVP) under Don Ostrander, an Air Force general formerly associated with the ARPA.[103] The Administrator and his deputies felt that booster development in general, and the Saturn program in particular, needed a more prominent position within the NASA organization. The new Marshall Space Flight Center with the von Braun team, the Saturn program, and NASA's facilities at Cape Canaveral (directed by one of von Braun's associates), became Ostrander's responsibility. Silverstein controlled Goddard, the Space Task Group (Project Mercury), and administered NASA's contract with the Jet Propulsion Laboratory. Wallops' position in this structure became a matter of debate.[104]

A 13 July 1959 Headquarters "Summary of Budget Policy Decisions" seemed to leave little doubt about Wallops. "It should be completely understood at all levels that the Wallops Station is an operational service facility and that engineering development programs are not a part of the station's mission. In view of the fact that Wallops is a service installation under the management of Space Flight Development, it is NASA policy that requirements for Wallops operation and support from A&SR or from outside NASA will be made through the SFD headquarters channel." The assumption was made that, "after the first few test Scouts, Wallops will be responsible for the assemble check out and launch of Scout vehicles. Payload check out will be accomplished by the cognizant development group."[105]

When Ostrander organized OLVP in November, however, the imminent acquisition of an orbital capability at Wallops indicated to him that the Office having responsibility for such launch vehicles, as well as NASA launch operations at the Cape and coordination of NASA operations at the Pacific

Missile Range, should also take charge of NASA's other range. Silverstein disagreed, and a 22 December meeting between the two attempted to iron-out the question of Wallops' position. "DSFD made a strong plea that W.I. operations remain research oriented and, therefore, should remain under his cognizance. ... Agreement on the assignment and responsibility for W. I. was not completely agreed to without reservation by DVDO."[106]

Silverstein acquiesced on Scout and agreed that the Langley-developed booster, and the Goddard-developed Delta, would be transferred to the Marshall center after test flights were complete. Prior to this Scout was slated to go to Goddard after the fourth test flight. Additionally, he agreed that, "all sounding rocket developments [are] to be the responsibility of DVDO."[107] The research oriented Silverstein, formerly of Lewis Lab, suddenly had to contend with a challenge from the non-NACA portion of NASA interested more in development. Though consigning such developmental issues to Ostrander, Silverstein insisted that, "Sounding rocket launchings would remain the responsibility of [Goddard]."[108] He fought to keep Wallops, and utilized an interesting argument in the process. "Dr. Silverstein pointed out that the launch facilities are a small part of the Wallops installation and that the installation primarily exists as an instrumented range."[109] Wallops neighbors, anticipating another Cape Canaveral at the base, and many civilian researchers looking to Wallops for access to space, would have found that statement curious.

Wallops remained within the OSFP jurisdiction and continued the process of establishing an independent administration. The close ties with Langley continued throughout this transition, despite the situation of the "mother lab" in a separate NASA division. Appointment of budget, claims, and safety officers, as well as officers to certify various formal actions, took precedence during the transfer. Recently hired staff from the naval base filled many of these positions.[110] Wallops also received the manuals, rate schedules, forms, and other paper paraphernalia that fuel a bureaucracy. While the NASA employees of today may complain about the increase in their paperwork, such red tape did not spring into existence overnight. Administrative independence required the Wallops staff to execute procedures previously left to Langley.[111] The two staffs, used to working with each other, accomplished the transfer with a minimum of disruption to the research program. Langley Director H.J.E. Reid set the tone of relations in a memo to his former employee, Krieger. "Langley shares your desire, ..., to effect an orderly transitional period. ... We shall be pleased to render whatever support we can during this period and for as long as necessary to insure continuance of your programs in the most efficient manner possible."[112] To be sure, Wallops remained small enough to escape the full onslaught of bureaucracy, and informal lines of communication still provided a quick means of resolving problems in these pre-Apollo days. Issues regarding patent and legal councils, for example, continued to be handled at Langley as Krieger deemed the need for such services at Wallops insufficient to warrant full-time positions at the base.[113] In March 1964, five years after formal separation of the facilities, Langley agreed to assist Wallops in the evaluation of the proposals for automatic data processing equipment.[114]

Adding to the administrative jumble was the fact that Wallops received operational direction and funding from multiple sources. While Silverstein's OSFP provided nominal oversight of the base and its flight operations, the tracking and data relay functions came under the purview of Edmund Buckley. In 1959 Buckley moved from Langley's IRD to NASA Headquarters to oversee the important tracking and data acquisition function for the space agency. As Assistant Director of Flight Operations under Silverstein, Buckley became "the contact for Wallops" at Headquarters.[115] Buckley's office grew more autonomous until, in 1961, the Office of Tracking and Data Acquisition became an separate Headquarters division. All the while Buckley, ever "a great friend" of Wallops, watched over the radars and telemeters at the base, cleared their funding, defended them before Congress, and generally saw to it that the equipment was kept modern and fully utilized.[116]

This meant that the Wallops staff not only provided services to a plethora of customers, but also answered to multiple segments of the NASA hierarchy, just as they had previously answered to differing divisions within Langley. They prepared proposed budgets, and justifications, in components at the base, then worked with the appropriate Headquarters office to finalize a given component. Then Headquarters would clear the budget with the Executive branch Bureau of the Budget. Once approved by the President, the Headquarters staff went before Congress. In this early period, Wallops personnel rarely testified before Congress, hence the value of friends like Dryden and Buckley who did.[117] The loose informal procedure (at their level) gave Wallops flexibility in planning operations. This promoted efficiency but impressed some in the established bureaucracy as an undisciplined way to operate.

Executive and Legislative desire to reign in the NACA constituted one consideration during the planning for NASA.[118] The General Services Administration conducted a "preliminary review ... of stores operations," at several NASA installations, including Wallops, in July and August 1959.[119] In August Headquarters audited radar usage at the base, and in December surveyed use of overtime.[120] In February 1960, an inventory control conference was held and included Ames, Lewis, and Wallops, which became the first to enact NASA's inventory control system.[121] During the NACA era, Wallops would have taken little notice of such proceedings, leaving them to Langley. Now, independence required Wallops to deal with these affairs. The continued support of Langley, and the respect of highly placed "friends" helped Wallops through this hectic time.

Krieger and company needed all the support they could get. A flood of research projects, released and made respectable by Sputnik, began to pour into NASA, and many of them required the utilization of the Station. The customer base, and staff, were growing so quickly that by late 1960 former PARD Chief Shortal sent Langley engineers to Wallops to acquaint the personnel at the rocket range with, "the type of research being conducted by AMPD at Wallops."[122] Langley projects like Scout and Trailblazer were joined by projects from other sources. Goddard initiated its space science program, Project Mercury testing began, and NASA encouraged universities to utilize the range.[123] Military projects did not suddenly disappear with the formation of the civilian NASA. While the creation of ARPA and the general elevation of the country's missile research facilities served to make Wallops somewhat less vital to the armed forces, military programs continued to come to the base. In 1958 the blast loads program (which utilized large explosive charges to simulate atomic airbursts) started testing an upgraded facility, which commenced operations in 1960.[124] Preflight Jet ran tests of B-58 bomber models for the Air Force, and tests of the Navy's Polaris missile also required this facility, as well as both rocket model and helium gun tests.[125] Sonic boom research flights offshore used the new radar systems that were coming on line.[126]

During Congressional testimony, NASA Director of Business Administration, Albert Siepert, answered questions concerning possible adverse effects of the civilian program at Wallops on military testing by stating, "If they [the DOD] wish to use [Wallops] we would be happy to work out arrangements."[127] It would appear that NASA had plenty of practice making such arrangements. However, in November 1959 the Air Research and Development Command (ARDC, an Air Force organization) liaison at Langley recommended stationing a liaison at Wallops, and throughout this period the military continued to supply boosters and other equipment to the Station.[128]

Not all proposed projects actually flew from the base, though. Safety continued to be a prime consideration in spite of the rapid pace. A Lewis proposal to test an engine utilizing hydrogen-fluorine fuel at Wallops posed a serious danger due to the toxic nature of the chemicals. Lewis researchers felt using the isolated island minimized the hazards of investigating the potential of this engine, hazards which would require the complete evacuation of the island during each test and a westerly wind to carry the dangerous exhaust gasses offshore. Concerns voiced by Krieger, and increasing research into hydrogen-oxygen engines led Abe Silverstein to cancel these test plans.[129]

The culmination of Wallops expansion came with the launch of *Explorer IX* on 16 February 1961. An inflatable sphere designed to study atmospheric density, this satellite became the first to ride into orbit atop an all solid-fueled vehicle, and made Wallops the third U.S. range with an orbital capability.[130] Enlarged, modernized, and independent, Wallops, like the rest of NASA, entered the 1960's anticipating a bright future.

NOTES

1. Walter A. McDougall, *The Heavens and The Earth: A Political History of the Space Age* (New York: Basic Books, Inc., 1985), 60.

2. The Army Ballistic Missile Agency (ABMA, the von Braun team) placed high priority on orbital flight, and pursued Project Orbiter. See: Frederick I. Ordway and Mitchell R. Sharpe, *The Rocket Team* (New York: Thomas Y. Crowell, 1979), 374-76. The Naval Research Laboratory represented the U.S. in the IGY effort with their Project Vanguard; see Green and Lomask, 25-56. RAND Corporation, originally an Air Force study group that became an independent research corporation, produced a series of reports detailing differing types of satellites throughout the 1950's. See, for example, Merton E. Davis and William R. Harris, *RAND's Role in the Evolution of Balloon and Satellite Observation Systems and Related U.S. Space Technology* (Santa Monica: RAND Corp., 1988). Different NACA space aspirations will be discussed later in this chapter. It should be noted, however, that none of these projects created much interest outside of their own project offices.

3. Bulkeley, 67-9 for the Soviet nuclear program. Green and Lomask, 192 97; McDougall, 131; "Spinak, et al.," OHI, Tape la: 376; for professional reaction to Sputnik. The reaction of those in the business seems to have been more irritation and frustration than unanticipated, out-of-the-blue shock.

4. Not only the satellites went into orbit on the Russian shots, but also the spent final stage of the launch vehicles. Green and Lomask, 194; McDougall, 150.

5. Enid Curtis Bok Schoettle, "The Establishment of NASA," in *Knowledge and Power: Essays on Science and Government*, ed. Sanford A. Lakoff (New York: The Free Press, 1966), 190. See also, McDougall, 150-51. PSAC had been part of the Office of Defense Mobilization until the Sputniks caused Eisenhower to move it into the White House. Killian, from MIT, had headed a Technological Capabilities Panel for the Science Advisory Committee in 1954 (Roland, I: 280); and Holaday, "had been responsible for the missile programs of the Department of Defense, ... ," (Bulkeley, 10-11).

6. Schoettle, 191; McDougall, 138-48,

7. Schoettle, 185-86; McDougall, 149-56.

8. Schoettle, 222-23.

9. Green and Lomask, 203-12, for the Vanguard failure. McDougall, 153-64, for Eisenhower's reactions during this period.

10. Roland, I: 296-99; Schoettle, 193-99.

11. Schoettle, 199-212, covers the services' differing points of view. Ordway and Sharpe, 382-86, for *Explorer 1*. Von Braun's team lost the IGY assignment to the Navy's Vanguard team, but quietly continued working toward the goal of launching a satellite. After the Vanguard explosion, they received permission to proceed on the project. The term applications satellite refers to a spacecraft with a specific, non research mission, such as communications or navigation satellites. The Polaris program involves the development and deployment of a submarine-launched nuclear missile.

12. Schoettle, 228-29; McDougall, 173. For Eisenhower's Farewell Address, in which he warns against the "military industrial complex," see U.S., President, *Public Papers of the Presidents of the United States* (Washington, D.C.: Office of the Federal Register, National Archives and Records Administration, 1953-), Dwight D. Eisenhower, 1960 61, 1035-40.

13. Shortal, 523; Roland, I: 290.

14. Roland, I: 290.

15. Virginia P. Dawson, "The Push from Within; Lewis Research Center's Transition to Space," in *A Spacefaring Nation: Perspectives on American Space History and Policy*, ed. Martin J. Collins and Sylvia D. Fries, (Washington, D.C.: Smithsonian Institute Press, 1991), 168. See also: Roland, I: 252-54; Hansen, 376.

16. Schoettle, 219, for quotation. See also, Roland, I: 259 81, 290-92.

17. Dawson, 172-75.

18. Hansen, 367-81; Hartman, 266-70. A lifting body is an aerodynamic shape that integrates the fuselage and wings in the production of the lifting force; the Space Shuttle is an example of a lifting body.

19. "Spinak, et al.," OHI, Tape la: 380.

20. Robert R. Gilruth, "From Wallops Island to Project Mercury, 1945-1958: A Memoir," in *History of Rocketry and Astronautics*, American Astronautical Society History Series, vol. 7, part 2, ed. R. Cargill Hall (San Diego, CA.: American Astronautical Society, 1986), 462. Dr. Hansen also noted this quotation on page 373 of his work.

21. Minute of Meeting, NACA Committee on Aerodynamics, 18-20 November 1957, in folder "Preliminary Space Flight Expansion Notes," Floyd L. Thompson Papers, Langley Research Center Historical Archives. Quotation is on page 17. This record group is hereafter cited as FLT Papers. Loyd S. Swenson, Jr., James M. Grimwood, and Charles C. Alexander, *This New Ocean: A History of Project Mercury* (Washington, D.C.: NASA, 1966), 56, describes the Committee on Aerodynamics as "the most influential of NACA's various technical committees."

22. U.S., Congress, Senate, Committee on Armed Services, *Inquiry into Satellite and Missile Programs, Hearings before a subcommittee of the Senate Committee on Armed Services*, 85th Cong. 1st sess., 5711-25S, p. 127, for Doolittle's testimony. Memo for the Press, "The NACA's Pilotless Aircraft Research Station, Wallops Island, Virginia," 17 May 1951, in folder "Wallops and Related Material," in the Milton Ames Collection, Langley Research Center Historical Archives. This record group is hereafter cited as MA Collection.

23. Memorandum, Robert L. Krieger to Associate Administrator Langley, 12 December 1957, in folder "Wallops, January 55 - February 58," in RGA181-I(C). Robert Gutheim, Chair of the Committee, "repeatedly gave assurances that because of the 'well recognized uniqueness and importance of NACA's work' nothing would be done to interfere with the NACA." For other groups see: "Robbins," Oral History Interview, Tape la: 170.

24. Roland, I: 292, II: 458; Hansen, 376-77. There is a photograph of the Stever Committee on Wallops Island in Shortal, 592. The Committee inspected the base on 28 October 1958.

25. Report, "NACA Research into Space," 10 February 1958, in folder "NACA Documents - 1958," in box "Administrative History: Pre-NASA Documents, NACA/DOD," in NHO. Quotation is on page 6. This report is an extract from an originally classified report with the same name issued in December 1957. A copy of this earlier report is in folder "January - December 57," Wallops box #4, WFFRC. Other pertinent reports include: "A National Research Program for Space Technology," "Resolution on the Subject of Space Flight," (both in FLT Papers); "A Program for Expansion of NACA Research into Space Flight Technology with Estimates of the Staff and Facilities Required," (reprinted in Roland, II: 730-64); and Air Mail, Floyd L. Thompson to Paul Purser, 29 January 1958, "Contemplated Expansion of Langley and W.I. Facilities Related to Space Flight Research Programs," (in folder "January - May 58," Wallops box #4).

26. U.S., Congress, House, Select Committee on Astronautics and Space Exploration, *Hearings before the Select Committee on Astronautics and Space Exploration on H.R. 11881*, 85th. Cong., 2nd. sess., 5804-15H, p. 404. The chart on this page gives the 90% figure. This figure recurs throughout the period and was arrived at by classification of all rocket launches, regardless of mission, as "space related activity." See also, Shortal, 593.

27. Memorandum for the President, "Organization for Civil Space Programs," The President's Advisory Committee on Government Organization, 5 March 1958, reprinted in James R. Killian, *Sputnik. Scientists, and Eisenhower* (Cambridge, MA.: MIT Press, 1977), Appendix #3, quotation on page 281. See also, McDougal, 171.

28. Ibid.

29. Schoettle, 229-31; McDougal, 169-76. These special committees later became standing.

30. Schottle, 260-61. This work provides an in-depth examination of the passage of the Space Act. The Space Act is reprinted in Robert L. Rosholt, *An Administrative History of NASA. 1958-1963* (Washington, D.C.: NASA, 1966), Appendix A.

31. Rosholt, 40-48; Roland, I: 299-300; McCurdy, 11-25.

32. T. Keith Glennan, *The First Years of the National Aeronautics and Space Administration: Events and Impressions as Recalled by T. Keith Glennan. 1st Administrator of NASA* (Cleveland: 1964), I: 6, in folder 002967 "Glennan Diary (1958-January 1960)," in file "Administrators, Glennan," in NHO.

33. Rosholt, 39-42, 48-50. Copies of both the Abbott Committee and McKinsey & Co. reports are in the NASA History Office. Neither directly discusses Wallops.

34. Rosholt, 33, for rubber-stamp criticism. It would be interesting to determine whose copy of the McKinsey Report found its way into NASA's archive. There are many, often critical, handwritten marginal notes apparently written by whomever reviewed the report, one of which states, "This is text book stuff - and not worth paying McKinsey to summarize for us." (Opposite page 2-2).

35. Shortal, 615. Also, charts in the Abbott and McKinsey reports depict Wallops under the space center's jurisdiction. See also, Rosholt, 48-50.

36. "Robbins," OHI, Tape la: 190; Shortal, 616-17.

37. Rosholt, 334-37, reprints eight organizational charts dated 18 July 1958 to 23 March 1959, all of which show Wallops listed under the space projects center. The final two charts, dated 29 January and 23 March 1959, also place operations at the Cape under this center (at this time NASA rented facilities at the Cape from the Air Force). The original charts are in NHO.

38. "Spinak, et al.," OHI, Tape la: 497, statement made by Abraham Spinak. See also Rosholt, 54-55. The Space Task Group will be discussed in further detail in Chapter 3, below.

39. Letter, G. E. MacVeigh to R. W. Hooker, 2 April 1959, requesting "various prints of different Wallops Island activities," so that the Beltsville Space Center could "plan our future activities." In folder "Special Files, April - June 59," in RGA181-l(S).

40. Minutes of Session IV, NASA Staff Conference 2-5 April 1959, in box "NASA Staff Conferences," in NHO. The list of attendees at this conference includes Reid and Thompson from Langley, DeFrance from Ames, Sharp from Lewis, Williams from HSFS, Pickering from JPL, and the Assistant Secretary of the Air Force for Research and Development (Horner). None of the staff from Wallops were invited despite the fact that the meeting was held in Williamsburg, Virginia.

41. Rosholt, 338, for reprint of chart dated 1 May 1959.

42. "Spinak, et al.," OHI, Tape la: 415-580; "Robbins," OHI, Tape la: 182 390; Shortal, 597-615.

43. Rosholt, 44, for observation about NASA employees. The launch log consists of two ledger books found in the Wallops Public Information Office. The entries are handwritten by John (Jack) Palmer.

44. Shortal, 596. Supporting documentation can be found in folder VIII "NASA Budget: FY 1959, FY 1960," in file tray "Budget, General Policy Planning, Chronologic, miscellaneous, FY 1959, FY 1960, FY 1961," in NHO. This file tray hereafter cited as "file tray NASA Budget." See also: U.S., Congress, Senate, Special Committee on Space and Astronautics, *Construction of Aeronautical and Space Research Facilities by the National Aeronautics and Space Administration*, S. Rept. 2076 to accompany H.R. 13450, 85th Cong., 2nd. sess., 5807-31S.

45. U.S., Congress, House, Committee on Appropriations, *Second Supplemental Appropriations Bill for 1958. Hearings before a subcommittee of the Committee on Appropriations*, 85th Cong., 2nd sess., 5801-27H. On page 138-9 Dryden notes that prior to Sputnik the military services had been providing motors to Wallops "without reimbursement." However, just when PARD needed more motors, "an internal accounting system in Defense," made it necessary for NACA to pay for the rockets.

46. McDougall, 201, for quotation from Glennan's diary. U.S., Congress, Senate, Committee on Appropriations, *Independent Offices Appropriations, for Fiscal Year 1959. Hearings before a subcommittee of the Committee on Appropriations on H.R. 11574*, 85th Cong., 2nd sess., 5804-30S, 170-79.

47. McDougall, 195-206. For concerns about NACA's creation see Roland, I: 5-25.

48. U.S., Congress, House, Select Committee on Astronautics and Space Exploration, *Authorizing Construction for the National Aeronautics and Space Administration. Hearings before the House Select Committee on Astronautics and Space Exploration on H.R. 13619,* 85th Cong., 2nd. sess., 5808-OlH, 28-29; U.S., Congress, House, Committee on Science and Astronautics, *1961 NASA Authorization. Hearings before the House Committee on Science and Astronautics on H.R.10246,* 86th Cong., 2nd. sess., 6002-17H, 46-47, for examples of testimony regarding facilities duplication. Shortal, 147, for NACA refusal to establish a West Coast range.

49. U.S., Congress, Senate, Committee on Astronautics and Space Science, *Investigation of Governmental Organization for Space Activities, hearings before a Senate subcommittee of the Committee on Astronautics and Space Science,* S. Rept. 806, 86th Cong., 1st sess., 5908-25S, 310-11.

50. Ibid.

51. Stever Committee Working Paper, "Launching Sites for Space," 17 March 1958, in folder "NACA Committee on Space Technology Working Group Papers," in box "Administrative History: Stever Committee Report 1958/Minutes," in NHO. Note that the table of launch sites in this report does not include Wallops.

52. For evidence of the re-evaluation of Wallops' expansion see Glennan's testimony before the Senate Subcommittee on Government Organization for Space Activities, page 49 of 5803-24S as cited in note 49 above. See also: Shortal, 626.

53. Shortal, 573-80, for details of "Project Jason," Wallops' portion of the ARGUS program. The atomic explosives were launched from a ship at sea, and timed to explode so that readings could also be obtained from the orbiting *Explorer IV* satellite.

54. Shortal, 580.

55. "Robbins," OHI, Tape la: 508-30.

56. Memorandum, J. Thomas Markley to Associate Director Langley, 24 November 1958, in folder "Wallops, March - December 58," in RGA181-l(C).

57. Memorandum with attachments, J. Thomas Markley for the Record, 23 December 1958, in folder "Wallops, March - December 58," in RGA181-l(C). Se also in same folder, Memorandum, Clotaire Wood to Administrator, 18 December 1958.

58. Letter of Agreement, Wallops Station and New York Regional Center, 1 March 1959, in folder "Wallops, January - December 59," in RGA181-l(C). See also in this file letters dated 12 February, 12 June, and 9 July 1959 regarding this subject. For communications between the Navy and NASA on Wallops clearance see: Letter, Thomas B. Combs (Chief of Naval Operations) to NASA, 21 November 1958; Memorandum, J. Thomas Markley to Associate Director, 24 January 1959: Letter, Edmund C. Buckley to CNO, 20 March 1959; all in folder "Special File, January - March 1959," in RGA181-l(S). An interesting overview on the topic is found in Memorandum, H. R. Brockett for Files, 19 September 1960, in chronologic file "July - December 1960," in box "NASA Headquarters Organization, O.A.R.T. (cont.), O.T.D.A., O.S.C.," in NHO. Paragraph 5 states, "A final coordination procedure which Wallops handles is that of fallouts outside of U.S. control. Tab D Presents the procedure followed which is based on methods followed by the Russians. As can be seen, this has the sanction of the FAA and is followed by Wallops when it applies." This box in NHO is hereafter cited as "NASA HQ box #1."

59. See Buckley's letter to the CNO, cited Ibid.

60. There are a number of handwritten form letters dated 30 June 1958 granting the director, Langley Aeronautical Laboratory permission to conduct surveys on land near state highway 680, in Wallops box #4. For contact with local lawyers see: "Robbins," OHI, Tape la: 208.

61. "Robbins," OHI, Tape la: 208-40.

62. Letter, E. Almer Ames to Harry Flood Byrd, 31 July 1958; Letter, Harry F. Byrd to H.J.E. Reid, 5 August 1958; Letter with enclosure, H.J.E. Reid to Harry F. Byrd, 7 August 1958; in folder "Special File, May - August 1958," in RGA181-l(S). See also: Letter, J.F. Victory to Howard W. Smith, 12 September 1958; Letter, T. Keith Glennan to A. Willis Robertson, 24 September 1958; Letter, T. Keith Glennan to A. Willis Robertson, 7 October 1958,: Letter, T. Keith Glennan to Harry F. Byrd, 8 October 1958; in folder "Reading File, July - December 1958," in box "1958-59 Chron. Files," in NHO. "Wallops Island Project Opposed," *Norfolk-Virginia Pilot* #17, 24 September 1958. Not all correspondence was negative, one man wrote offering to sell property to NASA for housing at Wallops; Letter, Paul J. Sterling to NASA, 14 October 1958; Letter, W. Kemble Johnson to Paul J. Sterling, 17 October 1958; in folder "Wallops, March - December 58," in RGA181 l(C). "What About Wallops Island," *Worchester Democrat*, 18 September 1958. Memorandum, Robert L. Krieger for Associate Director, 26 September 1958, in folder "January - December 1958," in Wallops box #4.

63. "Robbins," OHI, Tape la: 208-40; Shortal, 617.

64. Copies of these letters are in folder "Reading File July - December 58," in box "1958-59 Chron. Files," in NHO. See also: Letter, T. Keith Glennan to U.S. Attorney General, 30 January 1959, in folder "Reading file January 1959," box "1958-59 Chron. Files," in NHO.

65. U.S. Navy, Bureau of Yards and Docks, *Building the Navy's Bases in World War II* (Washington, D.C.: U.S. Government Printing Office, 1947), 237-38. Paolo E. Coletta, ed., *United States Navy and Marine Corps Bases, Domestic* (West Port, CN.: Greenwood Press, 1977), 114-16. Shortal, 616.

66. U.S., Congress, Conference, *Military Construction Act of 1956, Conference Report to accompany H. R. 9893*, H. Rept. 2641, 84th Cong., 2nd sess., 5607-07C, p. 6.

67. Letter, H. F. McKay to H. J. E. Reid, 5 May 1958; Memorandum, Robert L. Krieger to Associate Director, 13 May 1958; Memorandum, H. J. E. Reid to NACA, 16 May 1958; Letter, Hugh L. Dryden to H. F. McKay, 18 June 1958; in folder "Wallops, March - December 58," in RGA181-l(C). The Moffett transfer was apparently unaffected by the denial; Hartman, 316.

68. U.S., Congress, Senate, Committee on Appropriations, *A Bill for Supplemental Appropriations for Fiscal Year 1958*, H.R. 9131, 85th Cong., 1st. sess., 5706-03S, p. 306, for funding deletion. Joseph Robbins indicated his belief that the Navy closed CNAS due to general budget cutbacks; "Robbins," OHI, Tape la: 240. See also, "Anchors Aweigh," *The Washinaton Post and Times Herald*, 14 January 1959, A 18.

69. Memorandum, J. Thomas Markley to M. E. Phillips, 17 April 1959; Memorandum, Robert L. Krieger to Associate Director Langley, 7 April 1959; for the effect of the CNAS closure on Wallops, both are in folder "Wallops, January - December 59," in RGA181-l(C). For Navy-NACA cooperation on range clearance see: Letter (originally classified), H. J. E. Reid to Cmdr. John F. Betak, 4 April 1958, in folder "Special File, March - April 58," in RGA181 l(S); Letter, John Betak to Langley Aeronautical Laboratory, 5 August 1958, in folder "January 57 to December 57," in "Wallops box #4." Commander Betak was the Naval member of the New York Regional Airspace Subcommittee. See also a series of letters concerning meetings of this subcommittee in folder "Wallops, March - December 58," in RGA181-l(C).

70. "Robbins," OHI, Tape la: 240-323.

71. *National Aeronautics and Space Act of 1958*, sec. 302, a. Letter, Thomas S. Gates, Jr. to T. Keith Glennan, 9 January 1959, in binder "December 58 January 59," in box "Administrator's Staff Meeting Minutes (October 58 December 59)," in NHO. Gates was Secretary of the Navy. This binder includes minutes of the meeting held 19 January 1959, and contains staff paper #59-4, "Acquisition of Chincoteague Naval Air Station." The box is hereafter cited as "Staff Meetings box."

72. See minutes of staff meeting cited Ibid. Letter, T. Keith Glennan to Thomas S. Gates, Jr., 25 February 1959; Letter, F. A. Bantz to T. Keith Glennan, 5 March 1959; in folder "Wallops, January - December 59," in RGA181 l(C). "License for Use of Real Property by other Federal Agencies," Noy(R) 65516, Department of the Navy to NASA, 11 June 1959, in folder "June December 59," in Wallops box #4.

73. Shortal, 617.

74. NASA Press Release, "NASA Will Acquire Chincoteague Naval Air Station," 24 January 1959, in folder OW-050000-02 "Wallops Island Flight Center (NASA)," in the Space History Collection, National Air and Space Museum. Copies of letters to seven of the twelve landowners who had received notification of NASA's intent to acquire portions of their land are in folder "Reading File January 1959," in box "1958 59 Chron. Files," in NHO. Not all of the landowners received the good news. NASA still needed some land on the mainland for the causeway and radar installations.

75. "Robbins," OHI, Tape la: 250.

76. Minutes of Administrator's Staff Meeting, "Report on the Status of Chincoteague Transfer," 6 February 1959, in binder "December 58 - June 59," in Staff Meeting box. See also; Memorandum, E. C. Buckley to A. F. Siepert, 2 March 1959, in folder "Wallops January 58 - December 59," in RGA181-l(C). In "Wallops Made Key Space Site," *The Washington Post and Times Herald*, 13 September 1958, D 2, Krieger explicitly stated, "It won't be another Cape Canaveral."

77. "Robbins," OHI, Tape la: 280; "Milliner," OHI, Tape la: 130.

78. Memorandum, Albert F. Siepert to Wallops Station, 19 May 1959, in folder "Wallops January 59 - December 59," in RGA181-l(C). Siepert was Director of Business Administration at NASA Headquarters. For the complement of CNAS see: Press Release, J. E. Robbins to George McMath, 26 January 1959, in folder "January - May 1959," in Wallops box #4.

79. Letter, James F. Gleason to Thomas F. Johnson, 27 March 1959, in folder "Special File, January - March 59," in RGA181-l(S). See also; "NASA Administrator's Progress Report, August 1959," Part 1, 16.2; "NASA APR, October, 1959," Part 1, 16.2; both in collection of "Administrator's Monthly Progress Reports," in NHO. Letter, Cecil A. Duverney to Samuel T. Daniels, 25 June 1959; Letter, Ross Clinchy to Paul G. Dembling, 16 July 1959; Letter, Paul G. Dembling to Joseph Robbins, 17 July 1959; all in folder "Special File, July - August 59," in RGA181-l(S). Cecil A. Duverney, M.D. was President of the Worchester County Civic League. Samuel T. Daniels was Executive Secretary of the Maryland Commission on Interracial Problems and Relations. Ross Clinchy was Executive Director of the President's Commission on Government Employment Policy (which became The President's Committee on Equal Employment Opportunity on 6 April 1961). Paul G. Dembling was the Employment Policy Officer at NASA Headquarters.

80. For NASA hiring plans at Wallops see Gleason's letter to Congressman Johnson as cited Ibid. For hiring criteria in use at Wallops during this time see; "Robbins," OHI, Tape la: 240-323. Without compilation of a more complete set of records, such as the demographics of CNAS under Navy management or testimony from the individual making the allegation, determination of the validity of the charge can not accurately be made. Doubtless, few if any African Americans were hired from CNAS, (none of the photos that I have examined from this period show any black researchers or employees). To explain this situation as resulting from either: a conscious policy of discrimination (of which I have no evidence); an unconscious practice of discrimination (possible, owing to the nature of race relations in the U.S. in general at the time); or simply the unavoidable consequence of using a limited personnel budget to hire specialists from a few qualified applicants (few of whom would likely have been other than white due to the aforementioned tenor of racial relations) is a judgement that I can not make based on my current information.

81. "NASA APR, November 1960," Part 1, 16.2. The APR appeared monthly from July 1959 to December 1963, and hereafter will be cited by the initials APR and the date of issue.

82. Shortal, 599, 622-23.

83. Memorandum, T. Keith Glennan to Robert L. Krieger, 1 May 1959, in folder 004681, file tray "Centers, Wallops Flight Facility," in NHO.

84. "Spinak, et al.," OHI, Tape la: 560; "Robbins," OHI, Tape la: 345. See also: Memorandum, H. J. E. Reid to Robert L. Krieger, 12 June 1959, in folder "Special Files, April - June 59," in RGA181-l(S).

85. Memorandum, H. J. E. Reid to Paul G. Dembling, 30 October 1958; Shortal, 624-31.

86. Memorandum, Robert L. Krieger to Ralph Ulmer, 21 July 1959; "Spinak, et al.," OHI, Tape la: 445-90; "Robbins," OHI, Tape la: 392-410.

87. Shortal, 624-28. The implementation of Project 2080 was delayed briefly due to the range clearance controversy. One of Edmund Buckley's assistants passed an informal note requesting permission to proceed on 2080 to Hugh Dryden during an administrative meeting. Dryden wrote, "Yes, HLD," on the note, and 2080 commenced. Shortal references this note, and Spinak produced a copy of it during our interview.

88. *Data Book I*, 168.

89. Shortal, 31-2, 71, 628, for causeway construction. Shortal notes that the causeway was constructed along the route chosen in 1945. See also; page 197 for the boat explosion, and 200 for the seaplane mishap (Engineer McGoogan was interviewed for this project).

90. U.S., Congress, House, Committee on Appropriations, *Independent Offices Appropriations for 1959, Hearings before a subcommittee of the House Committee on Appropriations*, 85th Cong., 2nd. sess., 5802-18H, 566-71. See also, Shortal, 624.

91. "Robbins," OHI, Tape lb: 150.

92. U.S., Congress, Senate, Committee on Aeronautical and Space Sciences, *NASA Authorization for Fiscal Year 1960. Hearings before a Subcommittee of the Senate Committee on Aeronautical and Space Sciences on S. 1582 and H.R. 7007*, 86th Cong., 1st sess., 5905-21S, 791-93, quotation is on page 792. See also: Memorandum, H. J. E. Reid to R. E. Ulmer, 18 March 1959, in folder "Wallops, January - December 59," in RGA181-l(C); U.S., Congress, Senate, Committee on Appropriations, *Supplemental Appropriation Bill for 1960, Hearings before a subcommittee of the Senate Committee on Appropriations on H.R. 7978*, 86th Cong., 1st sess., 5907-13S, 26-29, during this hearing Senator Robertson questions Glennan, Dryden, Siepert, and Budget Office Ulmer about the affect of NASA's program on the communities surrounding Wallops.

93. Letter, E. C. Buckley to A. F. Siepert, 2 March 1959, in folder "Wallops, January 58 - December 1959," in RGA181-l(C).

94. MIT's Lincoln Lab was founded in 1951. "Funded largely by the Air Force, the Laboratory's field of work was to be the scientific and technical problems of air defense, and particularly of radar." Bulkeley, 33. "Spandar" stood for SPace rANge raDAR, Letter, Floyd L. Thompson to Robert L. Krieger, 31 October 1962, in folder "Special File October 1962 - April 1963," in RGA181-l(S). For good insights into the radar and tracking function at Wallops see: Memorandum, John C. McFall, Jr. to Associate Director, 4 May 1965, in folder "Wallops, January to June 46 [sic]," in RGA181-l(C); Memorandum, Robert D. Briskman for the Record, 27 July 1960, in "Chron. File, July - December 60," in NASA HQ box #1; U.S., Congress, House, Committee on Appropriations, *Independent Offices Appropriations for 1962. Hearings before a subcommittee of the Senate Committee on Appropriations*, 87th Cong., 1st sess., 6104-19H, p. 1128-31.

95. Telemetry systems utilizing frequency modulation (FM) provided greater clarity than older systems using amplitude modulation (AM). High-gain systems allowed a large volume of data to be received, and digital recording systems began to replace old mechanical "syncro-data" recording systems allowing more data to be registered. Reduction of data took place at Langley during this initial period, with "computers" (personnel, usually female, who performed the tedious mathematics) using a mechanical calculator called a "Freeden" to convert the recorded information into usable form. The introduction of early electronic computers accelerated this process. "Spinak, et al.," OHI, Tape 1b: 290-420. See also: Hallion, 10-11; Hansen, 84, 207; for "computers."

96. Report, "NASA Staff Conference, Monterey, California, 3 5 March 1960," 61, in box "NASA Staff Conferences," in NHO. *Data Book I*, 484-5.

97. Shortal, 706.

98. Shortal, 706-9. Ordway and Sharpe state that Hermann Oberth, Konstantin Tsiolkovsky, and Robert Goddard (the Big Three of rocketry), "proved that solid-propelled rockets could never be used to reach outer space," 380-81.

99. Shortal, 707.

100. The Javelin was a four stage vehicle capable of reaching nearly 1000 miles altitude. A design that developed simultaneously, designated Journeyman, was canceled by the Air Force, then revived by Goddard, but saw only limited use at Wallops. Ibid., 702-4.

101. Linda Neuman Ezell, *NASA Historical Data Book. Volume II: Programs and Projects 1958-1968* (Washington, D.C.: NASA, 1988), 61-67. Cited hereafter as *Data Book II*. The four stage Scout as originally developed could place a 130 pound payload into a 320 mile high orbit. Used as a probe, it could lift a similar payload to a height of nearly 2000 miles; lighter payloads could be lifted higher still. Presidential Science Advisor George Kistiakowski noted in his diary that the Scout "vehicle program looked sensible enough, although it is obvious that nobody wanted to use the relatively cheap Scout and so we will probably have this vehicle developed but not used. ... When [NASA doesn't] have good ideas they build expensive equipment." George B. Kistiakowski, *A Scientist in The White House* (Cambridge, MA.: Harvard University Press, 1976), 110.

102. "Spinak, et al.," OHI, Tape 2b: 380-450; NASA Release #60-285, "Address by Maj. Gen. Don R. Ostrander, USAF," 25 October 1960, 6-7, in folder "Launch Vehicles, Office of," in box "NASA HQ. Organizations: O.L.V., O.A.S.T., O.A.R.T.," in NHO. This box hereafter cited as NASA HQ box #2.

103. Rosholt, 115-16.

104. Ibid., 223. See also: McCurdy, 17-20.

105. Headquarters Staff Report, "Summary of Budget Policy Decisions," undated (probably 13 July 1959), 3-4, in folder VIII "NASA Budget: FY 1959, FY 1960," in file tray NASA Budget, in NHO.

106. Memorandum, Abraham Hyatt for Files, 23 December 1959, 2, in folder "Launch Vehicles. Office of," in NASA HQ. box #2. DSFD refers to Director Space Flight Development (Silverstein), and DVDO is Director Vehicle Development and Operations (Ostrander).

107. Ibid., 3.

108. Memorandum, D. D. Wyatt for the Record, 23 December 1959, 2, in folder "Launch Vehicles, Office of," in NASA HQ box #2. This memo and the one cited Ibid., provide the interesting opportunity to examine the meeting from two differing points of view. Hyatt worked for Ostrander, and Wyatt worked for Silverstein.

109. Ibid., 2.

110. "Robbins," OHI, Tape 1a: 182-208; Memorandum, Robert L. Krieger to H. J. E. Reid, 3 June 1959, in folder "June - December 59," in Wallops box #4; Memorandum, H. J. E. Reid to NASA Headquarters, 4 June 1959, also in that folder.

111. For views of Wallops employees on paperwork details see: "Spinak, et al.," OHI, Tape la: 95-140; "Robbins," OHI, Tape la: 182, Ib: 289-326; "Milliner," OHI, Tape la: 230-70, 456 98. For views of selected NASA personnel on this subject see: McCurdy, 180, questions 13 and 15. For administrative details transferred to Wallops see: Memorandum, Joseph E. Robbins to Norwood Evans, 7 May 1959, in folder "Special File, April - June 59;" Memorandum, T. Melvin Butler to Joseph Robbins, 1 July 1959, in folder "Special File, July - August 59," both in RGA181-l(S).

112. Memorandum, H. J. E. Reid to Robert L. Krieger, 12 June 1959, in "Special File, April - June 59;" Memorandum, Joseph E. Robbins to Langley, 6 August 1959, in folder "Special File, July - August 59;" Report "Physical Inventory of Stock Stores," Langley to NASA Headquarters, 5 August 1960, in folder "Special File, May - August 60;" all in RGA181-l(S). Memorandum, J. E. Robbins to T. Melvin Butler, 26 March 1961, in folder "Wallops, January March 61," in RGA181-l(C).

113. "Robbins," OHI, Tape lb: 289. Memorandum, Robert L. Krieger to F. L. Thompson, 26 November 1962, in folder "Wallops, August 62 - February 63," in RGA181-l(C). Memorandum, Robert L. Krieger to F. L. Thompson, 10 April 1961, in folder "Special File, April - August 61," in RGA181 l(S).

114. Memorandum, A. D. Spinak to F. L. Thompson, 4 March 1964, in folder "Special File, March - December 64," in RGA181-l(S).

115. *Data Book II*, 522-24. Shortal, 615.

116. "Milliner," OHI, Tape la: 190, for quotation. For examples of Buckley's interest in Wallops see: Memorandum, C. R. Morrison to Assistant Director Space Flight Operations [Buckley], 5 May 1960; Memorandum, C. R. Morrison to Assistant Director, 17 June 1960, both in "Chron. File, January - June 1960," in box NASA HQ box #1. See also, Memorandum, E. C. Buckley to Directors, 19 July 1960, in folder "Special File, May - August 60," in RGA181-l(S).

117. "Spinak, et al.," OHI, Tape 2a: 105-74; "Robbins," OHI, Tape lb: 289-326.

118. Roland, I: 301-3.

119. "NASA APR, July 1959," Part IV, 5.1-2; "NASA APR, September, 1959," Part IV, 5.3, in NHO.

120. Letter, H. J. E. Reid to James B. McElroy, 24 August 1959, in folder "Special File, July - August 59," in RGA181 l(S). NASA Staff Paper, "NASA Overtime Policy," 1 December 1959, in binder "July - December 1959," in Staff Meeting box, in NHO.

121. "NASA APR, February 1960," Section BA-5, 5.1; "NASA APR, April 1960," Section BA-5, 5.1.

122. Memorandum, Joseph A. Shortal for Files, 10 January 1960 [sic], in folder "Wallops, January - March 61," in RGA181 l(C). AMPD was PARD's successor at Langley, Shortal stayed with aeronautical research thus partially explaining why his book ends at this point.

123. Trailblazer was a project designed to launch an artificial meteor. This would provide data on both heating characteristics of materials, and a radar data base for tracking incoming objects (like warheads) travelling at high speeds. ARPA funded this project. "Spinak, et al.," OHI, Tape 1b: 340; Shortal, 673 p For Goddard projects see: Letter, G. E. MacVeigh to R. W. Hooker, 2 April 1959, in folder "Special Files, April - June 59," in RGA181 1(S). Project Mercury and university projects will be dealt with in chapters 3 and 4, respectively.

124. Shortal, 432-5; Memorandum, Harold B. Pierce to Associate Director, 24 October 1958, in folder "Wallops, March - December 58," in RGA181-1(C).

125. Shortal, 668-9, for B-58 fuel tank tests; 557-61, for Polaris tests. Note that both Polaris and the B-58 were nuclear delivery systems.

126. "NASA APR, July 1959," Part II, 1.1.

127. U.S., Congress, House, Committee on Science and Astronautics, *Missile Development and Space Science. Hearings before the House Committee on Science and Astronautics*, 86th Cong., 1st sess., 5902-02H, 56.

128. Letter, Cecil B. Bailey to ARDC, Andrews Air Force Base, 13 November 1959. "Milliner," OHI, Tape 1a: 498-520. Shortal, 542-3.

129. "Milliner,"OHI, Tape 1b: 380; "Spinak, et al.," OHI, Tape 1a: 470; Shortal, 610-12.

130. "NASA APR, October 1960," Section SFP, 46.3; *Data Book II*, 239. Note: prior to launch *Explorer IX* was designated S 56a.

Chapter 3

PILOTED SPACE FLIGHT

The American response to the challenge proffered by the Sputniks consisted of more than just an increase in missile research funding and the promotion of space science. Wounded national pride required a more tangible response, as did the need to demonstrate to both allies and adversaries abroad U.S. ability to operate on the Cold War's new front. President Eisenhower recognized this when he indicated support for projects designed to assure, "the U.S. does not have to be ashamed no matter what other countries do."[1] While not prepared to underwrite an exorbitant space spectacular, he did agree to a modest project that, if successful, would restore faith in American technical prowess. The perceived Soviet advantage could be nullified by putting a piloted satellite into space. NASA's Project Mercury proceeded with this "unstated," but widely held hope.[2]

The possibility of human space flight intrigued differing groups within the aeronautical community, and studies had been underway for several years. Military planners looked on space as the ultimate "high ground" in the Cold War. Some reviewed the theoretical work of German researchers Eugen Sanger and Irene Bredt, who had proposed a piloted craft that would skim the top of the atmosphere and possess intercontinental range.[3] Others thought of military bases on the moon.[4] Aerodynamicists working on the X-15 for the NACA's hypersonic program considered the rocket plane a step on the road to piloted space flight. Flying as high as 62 miles, the X-15 furnished the opportunity to test many items required for space flight while earning several pilots their astronaut wings.[5] A follow-on to the X-15 held out the promise of orbital flight and many NACA engineers believed that a winged spacecraft would be the ticket to the new frontier.[6] Political need for speed in putting a human in space dictated a program that would fly soon, but the complexities of winged spacecraft would take time to solve.

In 1957 PARD's Maxime A. Faget began researching the potential of a simple ballistic design for space flight which drew heavily upon work already conducted concerning ICBM warhead design. Engineer H. Julian Allen, of Ames Lab, had demonstrated the effectiveness of a blunt shape in overcoming the problem of re-entry heating. The conical shape of a vehicle utilizing such a design also simplified the problem of aerodynamic stability.[7] Another fundamental problem involved the limitations on the weight-lifting capacity of available boosters. The Atlas ICBM, the most powerful U.S. booster at that time, could only lift a weight of about one ton to orbit. A ballistic capsule would have to be designed within this weight restriction.

On 24 January 1958 PARD submitted a confidential 10 page report, "A Proposed Simple Means for Manned Space-Flight Research," to Langley management. The proposal called for a series of sub-orbital missions to be launched from Wallops. A piloted capsule would be boosted to an altitude of 100 to 200 miles by a solid-fueled rocket, make a parachute-slowed splashdown in the Atlantic, and be refurbished and reused. The group suggested the rocket, a cluster of seven Sergeant motors, be fired in stages of four, two, and one, with the second and third stages to be fired "at the pilot's discretion." The program carried an estimated cost of $2.4 million and a preparation time of eighteen months.[8]

Faget presented this concept in a paper at what turned out to be the last NACA Conference on High-Speed Aerodynamics, held at Ames 18-20 March 1958.[9] Other papers presented discussed an Ames design for an orbiting lifting body, and Langley's plans for a winged design founded upon data derived from the X-15 program.[10] While not the only ballistic design under study (several industry designs had been discussed prior to the conference), and disparaged by some as a "stunt" and "undignified," PARD's design met the weight limits imposed by Atlas, and the time scale imposed by politics. "The choice involved considerations of weight, launch vehicle, reentry body design, and, to be honest, gut feelings."[11]

After approval of the NACA's assumption of the civilian space mission, Robert Gilruth, who had encouraged and assisted the ballistic capsule studies, went to Washington, and received the assignment to formulate a piloted space program. He assembled a small team that included engineers from Langley and Lewis Labs, and set to work.[12] Faget and Paul Purser were members of this team; understandably, as "although no official approval for the development of a manned capsule had been received, Faget was able to obtain the support of a large section of Langley through personal persuasion."[13] The NACA moved swiftly, motivated more by the military's strong push to monopolize human space flight than by the Soviets. In August 1958 President Eisenhower directed that this endeavor be carried out by the civilian NASA, and several ARPA members were then integrated into Gilruth's planning group.[14] After roughing-out the program and receiving approval to proceed, Gilruth returned to Langley and organized a group of researchers to execute the project. The group became known, in November 1958, as the Space Task Group (STG).

The original membership of the STG contained a hefty percentage of PARD veterans, fourteen out of a total of thirty-six from Langley. Ten other researchers came aboard from Lewis.[15] Organized as an autonomous division of Langley, and reporting directly to Headquarters, the STG found itself involved in a highly complex, and highly visible, undertaking. The fact that so many of the STG came from PARD, and the group's location at Langley, guaranteed Wallops a major role in Project Mercury. Indeed, a very early program outline placed, "extension of the Wallops Island capabilities," as the first step in, "successful completion of the program."[16] Wind tunnels at Langley, and at several universities, were utilized, but the range at Wallops offered a convenient place from which to conduct flight tests without unduly disturbing military schedules at the Cape. The smaller Wallops base also received less notoriety than its Florida counterpart, allowing for a test program more reminiscent of NACA programs. Minor setbacks could be corrected without excessive criticism.[17]

The very basic shape of the Mercury spacecraft had already been investigated, but the specific details of the design remained to be finalized and tested. Determination of the aerodynamic characteristics of differing shapes became one major area of research. Testing different materials for use on the craft's heat shield, and refining the parachute recovery system were others. A series of tests (started in the summer of 1958, before the official start of the project) involved dropping models of varying complexity from balloons, and later from C-130 aircraft to discover the motions of experimental shapes during descent.[18] These tests allowed engineers to study designs for both drogue and main parachutes, and gave radar and telemetry operators the opportunity to test equipment and hone their skills.[19]

Sounding rocket launches in support of Project Mercury included the launch of two-stage vehicles to determine the aerodynamic characteristics of models traveling at nearly Mach 3, and five-stage vehicles which provided data on aerodynamic heating. Given Wallops involvement in the hypersonics program, such test flights were relatively routine.[20] Also routine were tests in the Preflight Jet that investigated heat ablation characteristics of different materials. While Wallops' wind tunnel could not provide a full range of test data, information generated there added to data obtained from tunnels at Langley and outside NASA to fill out a complete picture. The testing may have been routine, but the goal of the tests, a human in space, sparked an enthusiasm for the project that pervaded the operations.[21]

As the pace of the project increased, administrative arrangements shifted to reflect early shuffling within NASA. On 26 January 1959 T. Keith Glennan formally designated Gilruth both an Assistant Director of the Beltsville Space Flight Center (Goddard), and Director of Project Mercury. "Mr. Gilruth will serve under the direction of and report to the Director of Space Flight Development, Dr. Abe Silverstein."[22] Glennan desired to bring all space related activities to one field center. Unfortunately, that center had yet to be built, so the STG (like Wallops) continued to rely on Langley for support.

Tests at Wallops continued, and evaluations of the launch escape system occupied a prominent place in the work. The explosion of large boosters remained a common occurrence at the Cape, so Mercury planners conceived a way to lift the capsule and its crew away from danger should an emergency arise during launch. The system consisted of a solid-fueled rocket attached to the top of the spacecraft by means of a tower. If needed, the system could pull the capsule high enough to allow the pilot to activate the craft's parachute recovery system, and the STG deemed it an essential feature. The first tests of the escape system involved simply firing a tower and test capsule from a platform on the shore. These "beach abort" launches, conducted with both makeshift and production motors, proved that the system would work from a standing start. Obtaining in-flight performance data required a more elaborate series of tests, however, including a booster large enough to launch the tower and capsule combination, and simulate the aerodynamic conditions the Atlas would produce.[23]

To provide a vehicle for this task, one that would be operable from Wallops, Langley modified the booster design earlier proposed by PARD by reducing the number of solid-fueled motors from seven to four. The resulting "Little Joe" booster could hurl a production Mercury capsule and tower to a height of 100 miles and simulate the Atlas well enough to provide valid data.[24] The Little Joe also provided the means to flight test other capsule systems, and test the reactions of biological specimens (including monkeys) to the mission environment.[25] While not capable of providing orbital velocity, the booster allowed the execution of many preliminary tests without interfering with operations at the Cape, or necessitating the use of more expensive boosters.

Not all of the tests succeeded, of course. The most disconcerting failure came with the first attempt to launch the Mercury-Little Joe combination. A

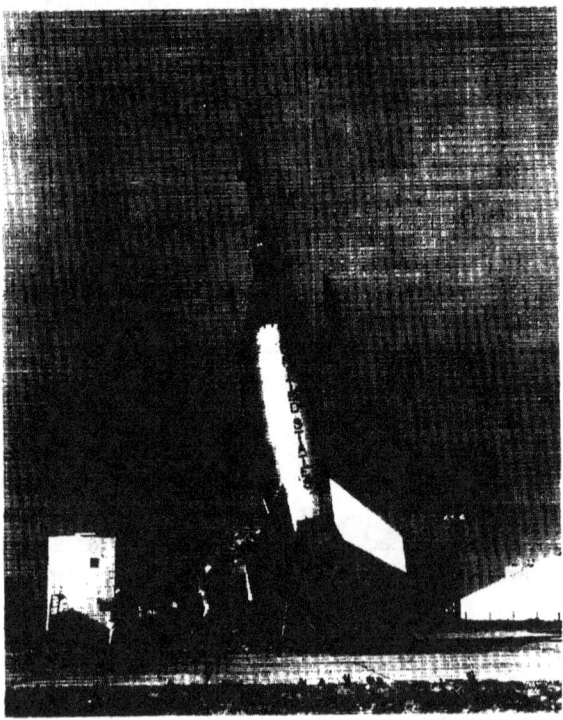

Little Joe vehicle with prototype Mercury capsule on Wallops launch pad.

short-circuit activated the abort system approximately thirty minutes prior to the scheduled launch. The escape tower pulled the capsule away from the booster, but the spacecraft's main parachute failed to deploy and the capsule was ruined.[26] On an earlier occasion a malfunction during a beach abort test caused a tower and capsule to somersault through the air and hit the water 1000 feet offshore.[27] The NACA-experienced Space Task Group recognized these failures as an inevitable part of a learning curve. This official tolerance for failure began to wane as NASA developed into a more visible enterprise. The intense negative public reaction to the Vanguard explosion was a harbinger of things to come.

Public relations constituted a novel trial for the space agency. "NACA was real low-key, they did not even have a public affairs officer, or division. ... they just didn't believe in public relations very much, ..."[28] Given that the data NACA generated dealt with highly technical engineering matters, a portion of which had military or proprietary value, this lack of official contact with the general public is understandable. Even the non-NACA portions of NASA (the Vanguard team, von Braun's group, and JPL) were more accustomed to military security than public relations. Wallops Station operated within this tradition. PARD once turned down the offer of a bore-sighted television camera as an accessory for their FPS-16 radar. "We said, we don't want that kind of stuff at Wallops, because we're not interested in showing this to the public. We thought that's what they were using it for at other places."[29]

The popular and politicized nature of the space effort did not give NASA the luxury of anonymity. The Space Act required the Administration to "provide for the widest practicable and appropriate dissemination of information concerning its activities and results thereof."[30] Though the public's attention focused on activities at Cape Canaveral, Wallops also attracted notice. These new responsibilities called for Wallops to expand this aspect of its operation as well.

A March 1958 request from the Department of Air Sciences at Maryland State College for "Static displays of a type that would tend to give the local civilian population an idea of your organization's function and relationship to the Eastern Shore," at a one day event had to be turned down. Wallops' limited supply of such materials was in use elsewhere.[31] In August 1959 a sounding rocket experiment that released a sodium-vapor cloud, "visible to ground observers within a 700-mile radius of the launch site," provoked a flurry of calls from citizens and officials startled by the strange apparition in the sky. The Wallops team had not foreseen this reaction.[32]

The situation began to improve as NASA's increased funding moved through the system. In April 1959 Headquarters directed Langley, "to have installed at Wallops Island twenty (20) telephones for use of the Public Information Office and press representatives."[33] A 30 June staff meeting at Headquarters discussed the matter of public relations during Project Mercury and concluded, "It will be his [the Director of the Headquarters Office of Public Information] to determine the propriety of releases, interviews, tours, and spot coverage of activities at the W.I. launching site, the A.M.R., and elsewhere." This served to coordinate public relations for Mercury and showed an awareness of the importance of this feature of the project.[34] Indeed, NASA's effort to disseminate information drew fire from Albert Thomas who berated Glennan and Dryden about the increased personnel cost this effort entailed, and the impression that this put "the pressure on the poor scientists," especially if a shot failed to launch. Glennan defended their public relations program as necessary to fulfill their legal requirements and pointed to Wallops as a place where publicity had been "controlled" (meaning, not excessive).[35] Thomas conceded the accuracy of this reference to Wallops, but Glennan still felt compelled to dispatch a memo to the center directors reinforcing NASA's launch policy. "I have stated many times, that the Test Conductor and Project Supervisor must understand that pressure from newspapers, distinguished visitors, or 'brass' from HQ. is not to influence a decision to attempt a launch. I want to reiterate my position on this matter."[36]

As Project Mercury progressed, public fascination with the space effort grew. The installation of bleachers on the mainland opposite the island provided ringside seats to rocket launches.[37] While the military generally exhibited a tolerant attitude toward security at Wallops, one military public information officer was somewhat taken aback at the sight of a grandstand full of people waiting to see a supposedly classified launch.[38] On one occasion it was decided to withdraw the base security guards for a weekend to demonstrate the open nature of the operation. However, "We had people wandering around where the rockets were stored, hitting on them, ... so ..., we said, you know, if they get blown up its going to be our fault."[39] The guards soon returned. That the experiment took place at all comes as something of a surprise given Krieger's feelings about safety. In an April 1960 memo he issued a strong warning about "unauthorized people on the island," brought to the launch site by employees traveling over the new causeway. "It is absolutely necessary that the entire island be considered an explosive area."[40] More official visitors also prompted Krieger to look to the appearance of his charge. "During the rapid growth of the island in area, new facilities, and the heavy workload, of all, it is apparent that we have neglected our policing and sight appearance on the island." Individuals were assigned to clean up various buildings and areas.[41]

The public interest peaked near the end of the Little Joe program when two flights carrying monkeys were launched. The first of these tests flew on 4 December 1959 and sent a number of specimens to an altitude of 53 miles. The second, on 21 January 1960, tested the monkey's reaction to the stress of a launch abort via the escape tower. Then Administrative Assistant Joyce Milliner recalled, "We had over a hundred photographers, I mean from every well-known news media," for the December launch. Even the stars of Mercury, the original astronauts, came to the island for the flight. While they were not to fly from Wallops themselves, nor participate in the test program there, they maintained a certain interest in the outcome of the test.[42]

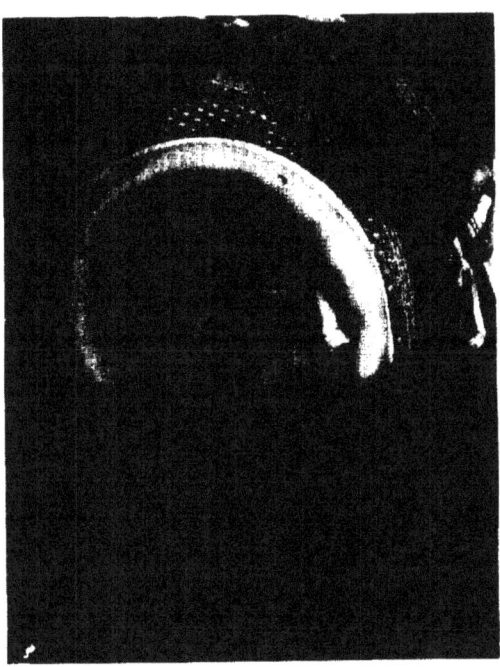

"Miss Sam" gazes from her contoured couch prior to flight test from Wallops in Little Joe 1B on January 21, 1960.

The final Little Joe flight launched on 28 April 1961, only one week before Alan B. Sheppard, Jr. flew Mercury-Redstone 3 to an altitude of 116 miles on a sub-orbital path to become America's first official space traveler. The disappointing part of the affair, for U.S. space enthusiasts, was the flight of Yuri A. Gagarin of the Soviet Union, who had completed one orbit around the Earth on 12 April. Round two of the space race went to the USSR.[43]

Wallops' role in Mercury did not end with the last flight of the Little Joe, however. Early in the planning process NASA realized that communications with a piloted vehicle represented a greater technical challenge than did communications with an automated satellite. Monitoring the pilot and the sophisticated craft during the experimental program necessitated high data transmission rates and nearly continuous contact. In 1956 construction had begun on the "Minitrack" network of tracking and data reception stations to support the Vanguard project. Though this network came to NASA with the rest of Vanguard, the limited range of operations and the restricted volume of data they could process made them unsuitable for Mercury.[44] The task of coordinating the development of a system capable of supporting the piloted flights fell on Edmund Buckley. In 1959 he commenced to work, first at Langley where he headed the Tracking and Ground Instrumentation Unit (TAGIU, a Langley group separate from the Space Task Group), an then from Headquarters as an Assistant Director in Silverstein's Office.[45]

Like many of facets of the space program, the design, construction, and operation of the "Manned Space Flight Tracking Network" (MSFN) departed from previous NACA practice and relied heavily on contractors. Engineers from Western Electric, Bendix, RCA, IBM, and others, worked with experts from Goddard and Langley to prepare the network.[46] The desire to beat the Russians to the punch did not leave NASA time to effect a slow build-up of capabilities and equipment from within, and the avowed civilian nature of the program precluded excessive military involvement. NASA needed help fast and turned to the private sector to obtain it.[47]

Additionally, the need for frequent ground to space contact dictated that NASA make arrangements for a global network that included specially equipped ships, and stations located on foreign soil. While most of the eleven Minitrack stations were situated outside the U.S., only one operated outside the Western Hemisphere. Placement of most of them along a rough "fence" running north to south provided contact with an orbiting satellite at least once each orbit so that the spacecraft could downlink information or receive instructions.[48] Mercury required a network encircling the globe in an equatorial fashion. Ships could cover open stretches of ocean, but NASA began negotiations with foreign governments to obtain permission to locate land stations where required. An important aspect of these negotiations rested on the civilian nature of NASA and Project Mercury. Many countries could not politically accept U.S. military bases on their soil. If the proposed NASA tracking stations were perceived to carry a military stigma, the negotiations would have rapidly broken down.

These departures from previous custom affected Wallops in several ways once the decision was made to locate the prototype Manual SpaceFlight Network station at the island. Buckley needed a place where research, development, and testing of the new systems could be performed. With Goddard still under construction, TAGIU at Langley, and a wealth of radar and telemetry experience and equipment already in place at Wallops, the base offered a convenient site for the "Evaluation/Training" facility. It also provided a place where visiting diplomats could observe the type of equipment and operations being discussed, and a politically safe location to train foreign personnel while allowing NASA to emphasize the civilian and scientific character of the program.[49] This decision brought a steady increase in the population of research contractors working at the base, and brought in foreign nationals who came to inspect and learn how to operate the equipment to be used in their countries.[50] It also resulted in Wallops' assumption of responsibility for a facility outside its own fences.

In March 1959 NASA received a request from Collins Radio Company to establish a communications link between Wallops and Bermuda.[51] Buckley, Krieger, and the other Wallops planners recognized that the increased range of the vehicles launched from the island, created a need for a tracking facility beyond the bounds of the existing installation. Scout, particularly, would require a downrange station as orbital insertion of the vehicle's payload would occur some distance away. In May 1959 NASA justified the Bermuda Station before Congress by pointing out that it would be used for both Mercury flights from the Cape, and Scout flights from Wallops.[52] On 13 November James F. McNulty of the STG was "authorized to proceed to Bermuda to initiate and supervise the construction of the Project Mercury station," to be built by Western Electric.[53]

Operational responsibility for the tracking station had yet to be decided upon, however. Buckley suggested either operating the facility with civil servants or contractors reporting to Krieger; contractors reporting to Walter Williams at Headquarters; or, "of course, the undesirable [option] of the contractor reporting to General Yates [Maj. Gen. Donald N., AMR Commander]. ... We have to decide it quickly to prevent Yates or someone else deciding it for us and also because the training of a few members of these crews start in the near future. ... I don't think we ought to wait on this."[54] NASA, still engaged in fighting for a niche in the federal bureaucracy, definitely wanted to limit military encroachment on its operations.[55] The agreement reached in a meeting with Assistant Director Hartley A. Soule at Headquarters allowed Western Electric to operate the station under Wallops' supervision until 30 June 1961, when the situation would be re-evaluated.[56] With a full flight schedule of his own to deal with, Krieger could not afford to dispatch scarce personnel to operate the station directly. The personnel involved with the MSFN Evaluation / Training facility at Wallops, it should be noted, did not report to Krieger. Instead, they reported to TAGIU, first at Langley, later at Goddard. Administering the contract and supplying flight controllers for non-Mercury launches constituted as much as Krieger could handle with his limited personnel budget.[57]

As the pace of several projects accelerated, the work-force shortage became more acute. In July 1960 a proposal surfaced to move the Blossom Point Minitrack Station from Maryland to Wallops.[58] The proposal was made despite a trip to Wallops, at which, "A discussion was held on the Wallops manpower situation. Apparently not only is there a serious shortage of personnel space, but there is also a severe limit on overtime. It was stated that Wallops had no idea how to man the NASA radar [Spandar], although a contract for this to an industrial organization was being considered. Also the amount of support from Langley IRD was in question. Wallops would like more IRD support, while Langley may tend to concentrate on only those Wallops projects that are of direct interest to IRD."[59] In spite of the need, after reviewing NASA's budget request for fiscal year 1961, Thomas'

Subcommittee, "denied salaries for 373 of the requested 962 new employees .... The majority of the increase in staff being requested in 1961 is required for the Goddard center and the Wallops station, where the impact of new and expanding duties and responsibilities is most urgently felt."[60]

Wallops complement did increase during this period (see appendix 3), but the commensurate increase in work-load stressed the Station's capabilities. In a communication to Shortal, Krieger related the situation during the summer of 1960: "Of course the trouble is that you guys, with a 3,000-man organization behind you, can put together a three-shift operation when you want to, whereas Wallops simply does not have, for instance, three shifts of radar people. As a result, on anything like Scout when your people need radiation checks or telemetry checks, or command-destruct checks at 11 o'clock at night and 2 o'clock in the morning, and start wind weighting at noon the next day to fire at 7 o'clock the following night, it means that my radar people do not have time between these various functions to go home and get a reasonable amount of sleep and get back. Although they may work only a matter of five or six hours out of 24, they cannot manage to get home for a period of 24 hours or more sometimes. The best we have been able to do so far to solve this problem is to put beds around the various places on the island and encourage our people during these two or three hour breaks to go climb in bed and get as much sleep as they can. This is not a solution at all, of course, but until I get more people and get some of them trained, I do not see how we can do any better."[61]

One labor problem Wallops mostly evaded centered on the difficulties experienced by other NASA, and Air Force, launch facilities with work stoppages. Strikes hampered construction efforts at operational missile bases as well as at test ranges. These job actions resulted in delays so severe that the Senate convened hearings on the issue. The Air Force estimated that by March 1961, 195 strikes at their 19 operational sites caused the loss of 50,500 worker-days, and 132 strikes at the three test sites lost 112,322 worker-days.[62] "Work stoppages in connection with organizational efforts and negotiations for new collective bargaining agreements account for more than half of the total man-days lost. A large portion of this occurred at Patrick [Cape Canaveral]. Jurisdictional disputes of all kinds are the second most important cause, accounting for more than a fourth of the total."[63] A June 1961 NASA staff meeting reported that, "The President has issued an executive order establishing an 11-man commission ... to deal with labor disputes at the three launch facilities, AMR, PMR, and Wallops. The Commission will have persuasive authority only, but the executive order has the effect of enforced arbitration."[64]

Despite this reference to Wallops, there seems to be no evidence of labor problems of this nature at the Station. The only strike at Wallops noted during this era involved a three-day walkout by the security guards starting on 28 August 1963, for higher wages. After Wallops administrators explained that the contract for security services was let after a competitive bid (a bid over which they had no control), the strike ended.[65] The small size of the expansion of the facilities at the base, relative to the massive build-up at other sites, mitigated many of the factors causing the strife. The slow economy of the Eastern Shore also made striking an unattractive option.[66]

Generally speaking, relations with both research and maintenance contractors seem to have been smooth. "Once we got into the mode of contracting for these services, they became a part of our team and we just sort of thought of them as Wallops employees. ... They were local people and friends, ..."[67] To comply with federal regulations prohibiting "fraternization" between civil servants and contract personnel, separate jobs and facilities were supposed to be provided. Wallops had neither the time nor the resources to always comply with this though. "We brought the contractors in and they were assimilated with the civil servants, and we caught the devil for that several times along the way. So every now and then we'd have to isolate the contractor and give him a radar to operate."[68] "Until they got really formal with the [Inspector General] and we had to sort of abide by the rules, we were lax on that."[69] Efficient accomplishment of urgent tasks called for bending the rules.

The lack of problems with organized labor was fortunate, as the pace of action did not slow as Mercury testing gave way to the operational phase of that program. While the Mercury effort focused on support for the tracking network, after the final Little Joe tests research work at the base reverted to an emphasis on sounding rockets. Programs like Scout and Trailblazer mover to the forefront of the agenda. In October 1960, Preflight Jet, inactive since February, was formally deactivated, its personnel having already been re-assigned. Several new tunnels coming on line at Langley rendered the unit surplus and obsolete.[70] At the same time the Helium Gun, in storage for "several years," also went. Useful for transonic tests, the needs the hypersonic and space programs exceeded its ageing capabilities.[71]

As if the hectic pace of activities was not enough to try the endurance of the Wallops personnel, Mother Nature contributed more excitement. In mid-September 1961 the approach of Hurricane Ester compelled Krieger to report, "Now battening down station. ... Will keep you informed if we can."[72] Though Ester missed the island, many local residents evacuated to the safety of the main base. Krieger noted afterward, "Handling more than 2400 refugees in 755 automobiles very interesting exercise. Seems any town has certain number of aged, sick, pregnant, and otherwise incapacitated citizens requiring doctor's care. Ended up operating 26-bed hospital complete with doctor, six nurses, etc. No births, deaths, or injuries to report." The Station itself escaped with only minor damage.[73] The previous December, Technical Services Division Chief William Grant had organized a Damaged Control Branch in his division. This Branch prepared contingency plans and a team to deal with potential catastrophic emergencies.[74] Planning of this nature undoubtedly facilitated readying for, and recovering from, the hurricane. Ester proved to be only a dress rehearsal, however. On 7 March 1962 (Ash Wednesday) a severe winter storm pounded the East Coast, flooding Wallops and Chincoteague Islands. The Main Base again played host to numerous refugees. Far enough removed from the coast to escape the sea's assault, the old Navy base provided a haven where relief efforts could begin. The magnitude of the storm was such that, for the first few days, the NASA personnel were on their own in the humanitarian work. One remembered it "as the time 5000 people came to dinner, and went home two weeks later."[75] The leaders at Wallops divvied-up their responsibilities: Grant oversaw damage control efforts on the island; Robbins saw to the needs of the civilian population; and Krieger coordinated contacts with local political leaders and NASA.

The task of providing for the refugees called for fast action on Robbins' part. After contacting the Red Cross and the Army, he set his staff to registering everyone who came off the helicopters that were evacuating Chincoteague, "so we could catalog who was here and who wasn't here." Feeding several thousand people in a cafeteria facility designed for only a few hundred also provided a challenge.[76] "Joe Robbins went out and started telling these companies to bring us food. We didn't know where we were going to get the money to pay, but we'll pay you. We had no authority to do it from Headquarters. We got authority after the fact."[77] By providing a centralized location for the effort, and "taking charge," the Wallops personnel kept a bad situation from deteriorating into a total disaster.[78]

The damage to the launch facility was heavy, but could have been worse. The storm breached the recently authorized, but incomplete, section of the seawall, flooding the island. The older section of the seawall had been stressed to the limit. The Ground Blast Apparatus was damaged so severely Langley decided not to rebuild it. The storm also completely destroyed Goddard's DOVAP radar facility, "including the steel mat approach road."[79] Sand clogged the underground infrastructure and lay several feet deep on the launch pads. Equipment and electronics had been thoroughly immersed in salt water, and several buildings sustained structural damage, including one housing a number of rocket motors.[80]

Many installations escaped with only minor damage, though. Grant's team began cleaning equipment and digging out. The Scout complex, only slightly battered, was repaired quickly enough so that the launch of Scout 9 took place on 29 March, only three weeks after the storm. Similarly, most of the Station soon returned to normal operation.[81] An immediate transfer of funds into "Project 3512," provided $1 million to initiate repairs at the Station. An additional million was required to complete the repairs, add to the seawall, and replace equipment that failed prematurely due to exposure to the elements.[82] For their efforts after the storm, the staff of Wallops Station received a NASA Group Achievement Award.[83]

As the operations at Wallops began to recover from the effects of the Ash Wednesday Storm, decisions involving the post-Mercury direction of the piloted spaceflight program, began to affect the base. President Eisenhower had not authorized any piloted program to succeed Mercury, but after Gagarin's flight dashed American hopes of putting a human in space first, something more substantial became necessary. The new Kennedy Administration, picking up the pieces of the Bay-of-Pigs fiasco, turned to NASA to reassure nervous allies and demonstrate Yankee competence to critics foreign and domestic. The chosen course of action was a highly publicized goal to send an American expedition to the moon before the end of the decade. NASA rose to the challenge by elevating its lunar plans from a low-priority, futuristic aspiration to the high-priority Project Apollo.[84]

After President Kennedy's decision to expand the piloted program into a major national effort, NASA implemented plans to establish a field center dedicated to human operations in space. The Mercury team worked at Langley, reported to Goddard, and operated as a semi-autonomous organization within NASA, largely independent of any existing field center. Since aeronautics, Langley's specialty; space science, Goddard's responsibility; and piloted space flight all presented differing and unique problems, Headquarters felt that a field center devoted to the latter would minimize interference and facilitate oversight. Located in Houston, the Manned Spacecraft Center provided Gilruth and company a place to call home.[85]

Detailed planning for the complex lunar endeavor could not begin in earnest until Mercury answered some basic questions. Indeed, it soon became apparent that an intermediary program would be required prior to any attempt to reach the moon; thus came Mercury Mark II, or Project Gemini. Certain general aspects of Apollo were immediately evident, however, and some of them related to the role Wallops would play in the mission. A piloted lunar mission entailed the launch of large payloads. This meant large, liquid-fueled boosters beyond the capabilities planned for Wallops. While plans existed for the development of large, solid-fueled boosters (larger even than the current Space Shuttle's solid boosters) these plans operated under Air Force auspices until late 1963. NASA concentrated primarily on the liquid-fueled systems familiar to most of its people in the interest of saving time. Any decision to utilize large launchers of either stripe from Wallops would have required an additional expansion of facilities that a cost-conscious Congress, adamant about avoiding duplication of facilities, would not have funded.[86] More importantly, Wallops suffered from the physics of its geographic location. Apollo required a launch site within 28.5 degrees of the Earth's equator. Cape Canaveral barely satisfied this requirement; Wallops, situated just shy of 38 degrees north, simply sat too far out of range to launch a piloted lunar mission.[87]

But what of Apollo research and development? Even though no astronauts ascended into orbit from Wallops during Mercury, the base unquestionably played a valuable role in testing hardware and training support personnel. The more advanced Apollo hardware would obviously require testing no less rigorous and thorough, while new techniques were needed for the ambitious lunar mission. Plans coalesced and in the Spring of 1962 NASA announced that Apollo equipment testing and evaluation would take place, not at Wallops, but at the Army missile range at White Sands, New Mexico.[88] At first, one wonders why NASA eschewed using facilities totally under its own control, or why Congress permitted such an apparent duplication. Why not modify and reuse the Little Joe equipment and employ the experienced personnel at Wallops instead of constructing new facilities and training new people at White Sands?

As with most decisions of this type, several factors played a role. The first involved the desire by many, both inside and outside NASA, to change spacecraft recovery modes from the splashdown of Mercury's water landings to a touchdown on dry land for Apollo. A splashdown utilizing a large, expensive naval fleet and the danger of losing a small spacecraft in the vast ocean was deemed a necessary evil for Mercury because of the relative ease of designing a capsule capable of providing a survivable water landing, and the wide margin for targeting error contained in those miles of empty ocean. Political pressures left insufficient time to overcome the engineering and navigational problems associated with a dry touchdown; Mercury had to fly, and quickly, so mission planners went with a water-based recovery mode.

The Soviet Vostok capsules, however, came down on dry land. This gave the Russian hardware an aura of technical superiority, and provided ammunition for NASA's critics. If part of NASA's job consisted of advancing astronautics the way the NACA advanced aeronautics, then Apollo better be able to land in Kansas; especially since the need to best the Soviets propelled the program.[89]

The sight of all those recovery ships did not sit well with Congress, and the vision of an Apollo capsule laden with precious moonrocks, and possibly moonwalkers, going the way of Gus Grissom's sunken *Liberty Bell 7* sat even less well with NASA planners. Many techniques were proposed to provide a safe touchdown, from an inflatable paraglider to retro-rockets that would fire just prior to landing. One factor common to all these schemes was the need for open land in which to conduct tests. Wallops provided access to lots of salt water, but little open land.[90]

The other major factor in choosing White Sands lay with the relocation of Gilruth's team. Moving the STG to Houston simply made it more convenient and cost-effective to use the facilities in New Mexico instead of traveling to the Eastern Shore to conduct tests. White Sands, opened by the Army in 1945, conducted sounding rocket firings and provided the Army with the same general capability that Wallops provided the NACA and NASA. Additionally, most of the major Apollo contractors were located either in California or Louisiana, and White Sands would be more accessible to them as well.[91]

Unlike the Mercury spacecraft, Apollo needed a significant orbital maneuvering capability, requiring new test apparatus. These liquid-fueled engines and thrusters, especially the Service Module engine, remained out of Wallops purview. The final determinant came when wind tunnel tests at Langley indicated that the proposed Apollo Command Module and Launch Escape Tower combination would be aerodynamically unstable if flown on the Little Joe booster. A newly designed, somewhat larger booster designated "Little Joe II" met the test criteria, leaving the original Little Joe facilities at Wallops unusable without extensive modifications. Little Joe had come off the drawing boards at Langley, but most of its designers, hard at work on the Little Joe II, now labored in Houston. The combination of geographic locations, new designs, and desire for both a consolidated test facility and for solid ground for test articles to land upon, eliminated Wallops from the site selection process.[92]

Testing of flight hardware at Wallops for Project Gemini was limited to experiments relating to the flexible-wing landing concept. Designed by Francis M. Rogallo of Langley, and considered a prime contender for both Gemini and Apollo touchdown systems, the device resembled a modern hang-glider. A short series of launches from Wallops in late-1959 tested basic features of the system. These flights proved that a folded para-glider could successfully be ejected from a canister and deployed at supersonic speeds. Most testing of the "Rogallo Wing" occurred in wind tunnels and at the Flight Research Center (the old HSFS), where rudimentary piloted tests flew. Technical difficulties with the system, and serious cost increases in Project Gemini as a whole, led to the adoption of conventional parachutes to recover the new spacecraft.[93]

While flight hardware testing for the piloted program waned, operations at the MSFN Evaluation/Training facility resumed. Gemini called for orbital mission durations of up to fourteen days, the expected length of a lunar mission, and several new tracking and data relay stations were required to cover this increased time on orbit. Five new land bases and two new shipboard stations came on line to augment the former Mercury network, and replace two that went out of service.[94] The Wallops facility, deactivated in December 1960, was reactivated the following July and evaluated improved instrumentation and trained new crews.

Preparing the facility for the new program proceeded in two stages. First, new classrooms were built to accommodate more students, and new courses reflected the improved equipment and the experience garnered from Mercury. Secondly, the updated instruments were installed and tested. Operation of the facility continued throughout the retrofit period, and both phases of the upgrade were largely complete by November 1963.[95]

In the Fall of 1961 the Tracking and Ground Instrumentation Unit moved from Langley to Goddard. Headquarters felt that consolidation of two of NASA's three tracking networks at one center would promote economy and efficiency. Goddard, already home for the unpiloted satellite network, seemed a logical place to host the consolidation, more so than the aeronautically oriented Langley.[96] In 1964 NASA carried the consolidation one step further. "With the advent of the Apollo Program, it is apparent that the training area, established for the Mercury-Gemini programs, will be unable to cope with the increased demand of the more advanced program and its new equipments."[97] Therefore, NASA sought $356,000 to relocate the facility to Goddard. Harry Goett, Director of Goddard, explained to Albert Thomas, "In order to train the crews that go out to the worldwide stations, we send them down initially to Wallops and train them there in the operation before they go out. We figure that there will be a considerable cost saving if we move them nearer to Goddard."[98]

The completion of construction at Goddard made it feasible to take the training center there, and despite the improvement in conditions around the Delmarva, the infrastructure around Wallops remained limited. Trainees sent to Wallops for "Gemini Phase I Training," had to reserve rooms at the Lord Salisbury Motel, over forty miles away. Tourists filled the local motels, and "other NASA commitments," filled quarters on the base itself.[99] So, just as most piloted space flight testing left Wallops when the STG left Langley for Houston, the major portion of the Island's participation in the MSFN left when TAGIU moved to Goddard.[100]

After relocation of the training facility Wallops' relation to the piloted space flight program became one of occasional support. Tracking facilities at the range assisted in tracking Saturn I - Pegasus, and Saturn V test launches. Scout flights supported Apollo heat-shield materials testing.[101] Helicopters performed a series of model drops to test the possibility of using para-gliders to recover lifting bodies, a support for early Space Shuttle research.[102] Wallops personnel would travel to other facilities to assist with piloted shots.[103] As the piloted programs and the space flight centers matured, however, tasks that once came to Wallops by necessity and convenience, went to more specialized installations. The result was a certain increase in specialization at Wallops. The program became more focused on the conduct of economical space science, with occasional forays into aeronautical research. Legacies of Project Mercury: contractors, public relations, and foreign researchers, did not wane with the piloted programs at the Station, though. These influences grew stronger as Wallops moved through the final phase of the transition era, the phase that saw the base move from simply carrying out the programs of others to planning some of their own.

NOTES

1. McDougall, 202.

2. Ibid., 200.

3. Richard P. Hallion, "The Antecedents of the Space Shuttle," in *History of Rocketry and Astronautics, vol. 10, of the American Astronautical Society History Series*, ed. A. Ingemar Skoog, (San Diego: A.A.S., 1990), 228-31. See also, Hansen, 350.

4. William E. Burrows, "Securing the High Ground," *Air and Space / Smithsonian* (December 1993/January 1994): 64-69.

5. Hallion, *On the Frontier*, 106-29. Systems needed in space that were tested on the X-15 included: reaction control jets, heat resistant materials, and a pilot pressure suit.

6. Gilruth, "Memoir," 463-67; Hansen, 377-81. The X-15 follow-on, known as Project HYWARDS, was an Air Force proposal that NACA personnel studied.

7. Hansen, 349-50; Shortal, 675-7. Allen perceived that a blunt shape moving rapidly through the air would create a shock wave that would keep much of the generated heat away from the nose-cone.

8. Langley PARD, "A Proposed Simple Means for Manned Space-Flight Research," 24 January 1958, in folder "January - May 58," in Wallops box #4. Shortal credited PARD's Max Faget and Paul Purser with this "scheme," page 650.

9. Hansen, 377-81, The paper Faget read was co-authored by Benjamin J. Garland and James J. Buglia.

10. Ibid.

11. Gilruth, 465-67.

12. Ibid., 467; *Data Book II*, 102.

13. Shortal, 437. Recall that, at the time, Shortal, PARD Chief, was Faget's boss.

14. Gilruth, 469; *Data Book II*, 98-99.

15. Ezell, *Data Book II*, 102n lists the Langley members of the STG by their prior affiliation.

16. NACA-Langley, "Facilities required for the development and test of manned space vehicles," undated, in folder "Preliminary Space Expansion Notes," FLT Papers. Quote is from "Summary" on page 8.

17. "Robbins," OHI, Tape 1b: 50-80. Shortal, 646.

18. Shortal, 637-38, 643-46.

19. Ibid., 643-46. The drogue parachute was small, designed to stabilize the spacecraft rather than slow it. It also served to extract the larger main parachute from storage.

20. Ibid., 638, 660.

21. Ibid., 636, 647. "Spinak, et al.," OHI, Tape 1b: 560.

22. Memorandum from the Administrator, T. Keith Glennan, 26 January 1959, in folder 3390 "Goddard Space Flight Center," in file tray "Administrators, Glennan," in NHO. See also: "Spinak, et al.," OHI, Tape 1b: 560.

23. Shortal, 647-49.

24. Langley PARD, "Proposed Simple Means," as cited in note 8; *Data Book II*, 48-9; Shortal, 647-9.

25. Shortal, 199-200; "Robbins," OHI, Tape 1b: 20-80.

26. Shortal, 656; "Spinak, et al.," OHI, Tape 2a: 50.

27. Shortal, 647.

28. "Milliner," OHI, Tape 1a: 84. In mid-1957, the NACA received a request from the publication *Missiles and Rockets* for information pertaining to solid-fueled motors. Headquarters forwarded the request where H. Lee Dickinson, the "Employee Relations Officer," was directed to deal with it. Memorandum with enclosures, H. Lee Dickinson to Walter T. Bonny, 28 June 1957, in folder "Wallops and Related Materials," MA Collection.

29. "Spinak, et al.," OHI, Tape 2b: 450, speaker is Marvin McGoogan. A bore-sighted camera was designed to be mounted on the radar and aligned with the radar's dish so that controllers could see what they were tracking. McGoogan stated that the partial failure of the first Scout might have been averted if trackers could have seen that their radar readings were in error.

30. *National Aeronautics and Space Act of 1958*, sec. 203a, 3.

31. Letter, Lionel R. Booth to John C. Palmer, 19 March 1958; Letter, H. J. E. Reid to Maryland State College, 9 April 1958; both in folder "Special File, March - April 58," in RGA181-1(S). Note that this institution is now the University of Maryland, Eastern Shore.

32. Letter, Thomas S. Combs to Director, Langley Research Center, 25 August 1959, in folder "Special File, July - August 59," in RGA181-1(S). See also: U.S., Congress, House, Committee on Appropriations, *Supplemental NASA Appropriation for 1960, Hearings before a House subcommittee of the Committee on Appropriations*, 86th Cong., 2nd sess., 6002-01H, 16.

33. Memorandum, NASA Headquarters to Langley Research Center, 17 April 1959, in folder "Wallops, January - June 46 [sic]" in RGA181-1(C).

34. Minutes of Administrator's Staff Meeting of 30 June 1959, page 3 and annex 2, in binder "December 58 - June 59," in Staff Meetings box, NHO.

35. *Supplemental NASA Appropriation for 1960*, 16-17, as cited in note 32 above.

36. Memorandum from the Administrator, T. Keith Glennan, 10 March 1960, loose in binder "NASA Staff Conference, Monterey, CA., 3-5 March 1960," in box "NASA Staff Conferences," in NHO. It is interesting to note, that even after this admonition, during the Administrator's staff meeting of 3 November 1960, "The problem of scheduling launches in a highly charged political climate was discussed. Flight dates during this period, as at all other times, will of course be governed by technical considerations only." Minutes of Administrator's Staff Meeting of 3 November 1960, 2, in book #1, in box "Administrator's Staff Meeting Minutes, October 60 - June 61," in NHO.

37. "Milliner," OHI, Tape 1B: 120. NASA Wallops News Release, "Wallops Station Contract Awards During November," 3 December 1962, lists a contract for bleachers with Evans Construction Company in the amount of $26,400; in file tray "Wallops Flight Facility (cont.)," in NHO.

38. Ibid., Tape 1a: 350. The incident occurred in the late-60's or early 70's, but accurately reflects the situation at Wallops throughout the era. See also: "Spinak, et al.," OHI, Tape 1a: 140, for low level of security at the base.

39. "Robbins," OHI, Tape 1a: 535.

40. Memorandum, Robert L. Krieger to Staff, 14 April 1960 in folder "Special File, January - April 60," in RGA181-1(S).

41. Memorandum, Robert L. Krieger to Staff, 7 June 1960, in folder "Wallops, January - June 60," in RGA181-1(C).

42. "Milliner," OHI, Tape 1b: 120; "Spinak, et al.," OHI, Tape 2a: 25-50. For technical details of these flights see, Shortal, 656-9.

43. McDougall, 243. See also: *Data Book II*, 139-43 for tables 2-28 and 2-29.

44. *Data Book II*, 534-38; Rosholt, 45.

45. *Data Book II*, 521-24.

46. Ibid., 546-47. Old ways die hard, though. In February 1959 John Crowley sent a memo to Floyd Thompson at Langley: "It is suggested that Langley make use, wherever practical, of the personnel and facilities of the Pacific Missile Range, the White Sands Missile Range, Eglin Gulf Test Range, and the Atlantic Missile Range." Notice that he apparently felt it unnecessary to suggest using Wallops. Memorandum, John W. Crowley to Floyd L. Thompson, 20 February 1959, in folder "January - May 59," in Wallops box #4.

47. McCurdy, 34-40, 134-36; McDougall, 200; Rosholt, 156.

48. *Data Book II*, 534-36, 540. The 11 Minitrack stations included 3 in the U.S. (Blossom Pt., MD., Ft. Stewart, GA., San Diego, CA.); 3 Caribbean sites (Antigua, Grand Turk, Havana); 4 in South America (Anzofagasta and Santiago, Chile; Lima, Peru; and Quito, Ecuador); and 1 in Woomera, Australia.

49. U.S., Congress, House, Committee on Science and Astronautics, NASA *Authorization for 1965, Hearings before a House subcommittee of the Committee on Science and Astronautics on H.R. 9641*, 88th Cong. 2nd sess., 6402-18H, part 4, p. 2129. See also: *Data Book II*, 546. For the influence of foreign observers see: Memorandum, Edmund C. Buckley for the Record, 5 July 1960, in "Chron. File, July - December 60," in NASA HQ box #1. This memo describes the fourth meeting of "The Mexican - U.S. Commission for Space Tracking Observations - Project Mercury." On 27-30 June 1960, the Commission met at Wallops, toured the base, and viewed the equipment. Buckley was Chairman of the U.S. section of the Commission. See also: Letter, Edmund C. Buckley to General Antonio Perez-Marin, 14 December 1962, in "Chron. File, January - December 1962," in the same box. This letter describes the arrangements to be made for the training of Spanish personnel for the tracking station at Las Palmas (Grand Canary Island). That NASA could cooperate with Francisco Franco's government without drawing severe Congressional fire is probably a testament to the perceived importance of the space program.

50. "Spinak, et al.," OHI, Tape 2a: 545, notes that the influx of contractors was not abrupt. For foreign nationals see: U.S., Congress, House, Committee on Science and Astronautics, *NASA Authorization for 1964, Hearings before the House Committee on Science and Astronautics on H.R. 5466*, 88th Cong. 1st sess., 6303-12H, part 4, 2845. It should be noted that many of the foreign stations were operated by contract to Bendix, while employing local people.

51. Letter, Robert C. Miedke to G. B. Graves, 12 March 1959, in folder "Special File, January - March 59," in RGA181-1(S).

52. U.S., Congress, Senate, Committee on Aeronautical and Space Sciences, *NASA Authorization for 1960, Hearings before a Senate subcommittee of the Committee on Aeronautical and Space Sciences on S. 1582 and H.R. 7007*, 86th Cong. 1st sess., 5905 21S, 787.

53. Letter, E. C. Buckley to J. F. McNulty, 13 November 1959, in "Chron. File, January - June 1960," in NASA HQ box #1. This directive reminds McNulty that NASA's agreement with the Bermuda Government remained informal, and thus he was to keep a low profile while there and avoid the press, until a formal agreement could be signed.

54. Memorandum, E. C. Buckley for F. B. Smith, 16 November 1959, Ibid.; Walter Williams was STG's Associate Director for Operations at this time.

55. Arnold S. Levine, *Managing NASA in the Apollo Era* (Washington, D.C.: NASA, 1982), 211-37, for NASA - military relations.

56. Memorandum, Robert L. Krieger for Wallops Station Files, 15 March 1960; Memorandum, Hartley A. Soule for Files, 23 March 1960; both in folder "Special Files, January - April 60," in RGA181-1(S).

57. Ibid.; *Data Book II*, 545-47.

58. Memorandum, Robert D. Briskman for Assistant Director, Space Flight Operations, 13 July 1960, in "Chron. File, July - December, 1960," in NASA HQ box #1. The Blossom Point Minitrack Station was the prototype facility.

59. Memorandum, Robert D. Briskman for the Record, 13 July 1960, Ibid. Quote is on page 2. Notice that Briskman was also the person recommending placing the Blossom Point Station at Wallops, even after seeing the situation firsthand. The Minitrack station eventually moved to Goddard.

60. U.S., Congress, Senate, Committee on Appropriations, *Independent Offices Appropriations for 1961, Hearings before a Senate subcommittee of the Committee on Appropriations on H.R. 11776*, 86th Cong, 2nd sess., 6005-16S, 238.

61. Letter, Bob Krieger to Joe Shortal, undated, attached to Memorandum, C. C. Shufflebarger for Associate Director, 1 August 1960, in folder "Wallops, July - August 60," in RGA181-1(C).

62. U.S., Congress, Senate, Committee on Government Operations, *Work Stoppages at Missile Bases, Hearings before a Senate subcommittee of the Committee on Government Operations pursuant to Senate Resolution 69*, 87th Cong. 1st sess., 6104-25S, part 1, 12. The three test sites, Patrick, Vandenberg, and Edwards Air Force Bases, all hosted NASA operations.

63. Ibid., 13 for quote, 250-51 for statistics.

64. Minutes of Administrator's Staff Meeting, 1 June 1961, p. 1, in book #2, in box "Administrator's Staff Meeting Minutes, October 60 - June 61," in NHO. One of the alleged reasons for the strikes was dissatisfaction with the Davis-Bacon Act of 1931 that established guidelines for compensation of labor used in government construction projects. While a review of the Act, conducted in 1963, concluded that the Act, "is badly in need of a general overhaul and updating," it also stated that "the allegation that the Davis-Bacon Act was the cause of work stoppages ... at missile bases, [was] not generally substantiated by other testimony received." Perhaps. Two years later, however, "Wallops Island is still concerned about labor rates under the Davis-Bacon Act and the classification of employees working on Wallops Island contracts." For the D-B Act see: U.S., Congress, House, Committee on Education and Labor, *Administration of the Davis-Bacon Act, Hearings before a House general subcommittee of the Committee on Education and Labor*, 88th Cong. 1st sess., 6306-00H, 13. For Wallops' concern see: Memorandum, J. B. Sollohub for File, 5 February 1965, in folder "Special File, January - May 65," in RGA181-1(S); quote is on page 3.

65. NASA Historical Staff, *Astronautics and Aeronautics, 1963* (Washington, D.C.: NASA, 1964), 325 (this annual chronology is hereafter cited in the form "*A&A, date*"); "Robbins," OHI, Tape 1b: 150. It is interesting to note that the company listed in Wallops' telephone directories during 1962 and 63 (Plant Security Inc.) as providing guard services, was replaced in the 1964 directory by a new entry (Metropolitan Security Services, Inc.), at the 2-319 phone number. Telephone directories are in box "Center Telephone Books; Stennis Space Center (1990 -); Wallops Flight Center (1962 - 1979)," in NHO. See also: "Floyd," OHI, Tape 1a: 450.

66. "Robbins," OHI, Tape 1b: 160-200, for Wallops' relative size in NASA. "Milliner," OHI, Tape 1a: 360, for economic conditions on the Eastern Shore.

67. "Milliner," OHI, Tape 1b: 152. It should be noted that contractors were classed as either "service" contractors, providing operational services (such as security or grounds maintenance); or "research" contractors, fulfilling program contracts for various segments of NASA (such as operating a radar). This distinction was emphasized during the course of each oral interview.

68. "Spinak, et al.," OHI, Tape 2a: 545.

69. "Milliner," OHI, Tape 1b: 152.

70. Memorandum, Robert L. Krieger to Langley, 6 October 1960; Memorandum, Floyd L. Thompson to Wallops, 1 November 1960, both in folder "Wallops, September - November 61," in RGA181-1(C), (note that they are misfiled). For tunnels coming on line at Langley during this time see *Data Book I*, 356-7. The deactivation of Preflight Jet at this time is somewhat ironic in view of a strongly worded memo written by Krieger in August 1957 in response to a "ridiculous" proposal to relocate the wind tunnel to Langley. Memorandum, Robert L. Krieger for Associate Director, Langley, 6 August 1957, in folder "Wallops, January 55 - February 58," in RGA181-1(C).

71. Memorandum, Wallops Station to Langley, 6 October 1960, in folder "Wallops, September - December 60," in RGA181-1(C).

72. Teletype, NASA Wallops Station VA to Langley Research Center - Director, 19 September 1961, in folder "Wallops, September - November 61," in RGA181-1(C).

73. Teletype, NASA Wallops Station VA to NASA HQ - DR Abe Silverstein, 21 September 1961, Ibid. See also: "NASA APR, September 1961," section SFP, 27.4. "The Main Base," or simply "the Base," usually refers to the old Navy base in the lexicon of the Wallops Station.

74. Memorandum, William E. Grant for All Concerned, 16 December 1960, in folder "Wallops, September - December 60," in RGA181-1(C).

75. "Robbins," OHI, Tape 1b: 216-275. Quote is near 220.

76. Ibid.

77. "Milliner," OHI, Tape 1b: 310-80, quote is near 340.

78. Ibid. See also: W. Corlett Galvin, "Report of Storm Damage Suffered by Wallops Island During the Period March 6-9, 1962," p. 3, in "Chron. File, January - December 1962," in NASA HQ box #1.

79. W. Corlett Galvin, "Report," Ibid. Memorandum with attachments, Floyd L. Thompson for NASA - Code RTM, 5 April 1962; Teletype, NASA LRC Langley AFB VA to NASA HQ Wash D C, 11 March 1962, both in folder "Special Files, January - April 62," in RGA181-1(S). Teletype, NASA Wallops Station VA to NASA Headquarters, 12 March 1962, in folder "Wallops, January - March 62," in RGA181-1(C). DOVAP stands for DOppler Velocity And Position. Shortal's work ends prior to the Ash Wednesday Storm, but on page 630 there is a picture of the breached seawall.

80. Memorandum, N. Pozinski for Files, 9 March 1962, in "Chron. File, January - December 62," in NASA HQ box #1.

81. "NASA APR, March 1962," page D 2.4, for Scout 9. "Spinak, et al.," OHI, Tape 1b: 430-75.

82. Memorandum, Edmund C. Buckley to Director, Office of Programs, 15 March 1962; Report, "NASA Project No. 3512: Wallops Station, Storm Damage Repairs, undated; Memorandum, W. C. Galvin for Files, 21 March 1962; Report, "Cost Breakdown on Instrumentation Damage at Wallops Island, undated; Memorandum, W. Corlett Galvin for File, 2 May 1962; Memorandum, Edmund C. Buckley for the Associate Administrator, 17 May 1962; Memorandum, W. Corlett Galvin for the Record, 22 May 1962; Letter, Milton E. Stevens to G. S. Brown, 5 April 1962; all in "Chron. File, January - December 1962," in NASA HQ box #1.

83. Memorandum, Edmund C. Buckley for the Associate Administrator, 12 March 1962; Letter, James E. Webb to Robert L. Krieger, 13 March 1962; Memorandum, Edmund C. Buckley to the Associate Administrator, 21 March 1962; Memorandum, Robert C. Seamens, Jr. to Patrick Gavin, undated; all in "Chron. File, January - December 62," in NASA HQ box #1. See also: Letter, Robert L. Krieger to Floyd L. Thompson, 2 July 1962, in folder "Special File, May - September 62," in RGA181-1(S); and "Wallops Station Staff To Receive NASA Award," *The Salisbury Times*, 21 June 1962.

84. John M. Logsdon, *The Decision to Go to the Moon: Project Apollo and the National Interest* (Cambridge, MA.: The MIT Press, 1970), 100-30. U.S., President, *Public Papers of the Presidents of the United States* (Washington, D.C.: Office of the *Federal Register*, National Archives and Records Administration, 1953 -), John F. Kennedy, 1961, pp. 396-406.

85. Rosholt, 213-14; Levine, 19, 33; Data Book I, 390-94.

86. Levine, 227-28 for the advanced solid motor project. This motor measured 21.6 feet in diameter as compared to the Shuttle SRB's 12.1 feet. See also: U.S., Congress, House, Committee on Science and Astronautics, *Development of Large Solid Propellant Boosters, Hearings before a House Special subcommittee of the Committee on Science and Astronautics*, 87th Cong. 2nd sess., 6208-08H. No mention of Wallops appears in this record.

87. NACA Report, "Launching Sites for Space," 17 March 1958, 1-3, in folder "NACA Committee on Space Technology Working Group Papers," in box "Administrative History, Stever Committee Report 1958 / Minutes," in NHO. See also: James E. Oberg, "Rendezvous in Space," in *Air and Space / Smithsonian* (August/ September 1993), 44.

88. Courtney G. Brooks, James M. Grimwood, and Loyd S. Swenson, Jr., *Chariots for Apollo: A History of Manned Lunar Spacecraft* (Washington, D.C.: NASA, 1979), 91 92.

89. Memorandum, Robert F. Freitag to Distribution List, 27 February 1964; Memorandum, Robert F. Freitag to Distribution, 5 March 1964; both in folder 007153 "Land Landing Files," in collection "Project Apollo," in NHO. See also: U.S., Congress, Senate, Committee on Aeronautical and Space Sciences, *Manned Space Flight Programs of the National Aeronautics and Space Administration: Projects Mercury, Gemini, and Apollo*, S. Staff Report, 6209-04S. A drawing on page 149 depicts an Apollo capsule touching down on land. In truth, the Vostok capsule could not provide a "survivable" landing. Gagarin and his fellow Vostok cosmonauts left their ships after re-entry and parachuted to Earth separately. For reasons of propaganda and prestige the Soviets could not officially admit this fact, and it took awhile for the nature of the Vostok landing system to become widely known in U.S. policy circles. When the fact did become known, it combined with the unforeseen cost and complexity involved in designing a touchdown system, and the desire not to lose the (perceived) close lunar race over the issue, to quiet demand for the system. Apollo flights, like Mercury and Gemini before them, ended with a splash. James E. Oberg, *Red Star in Orbit* (New York: Random House, 1981), 54-5; *A&A, 1963*, 268.

90. Grissom's Mercury capsule sank in the Atlantic after landing when the hatch prematurely jettisoned; Grissom almost went with it. *Data Book II*, 144. For alternate landing methods see: Memorandum, Edward Z. Gray to Associate Administrator for Manned Space Flight, 18 August 1964. "Apollo Testing to Begin at White Sands," *Aviation Week and Space Technology*, 22 July 1963, 281. U.S., Congress, Senate, Committee on Aeronautical and Space Sciences, *NASA Authorization for Fiscal Year 1964, Hearings before the Senate Committee on Aeronautical and Space Sciences*, 88th Cong. 1st sess., 6306-12S, 1045-45.

91. Rosholt, 213-14; Levine, 19, 33; *Data Book I*, 395-97.

92. Ivan D. Ertel and Mary Louise Morse, *The Apollo Spacecraft: A Chronology, Vol. I,* (Washington, D.C.: NASA, 1969), 141. NASA News Release, "Apollo at White Sands, by J. Thomas Markley," 11 September 1962, in folder 007215, in file tray "XII Manned Space Flight: Apollo - Little Joe II," in NHO. Interestingly Wallops had borrowed some equipment for Little Joe from White Sands. See: Letter, R. L. Barber to Commanding General, White Sands, 22 May 1959, in folder "Special Files, April - June 59," in RGA181-1(S); and Memorandum, E. C. Buckley to Director, Space Flight Programs, 31 October 1960, in "Chron. File, July - December 1960," in NASA HQ box #1.

93. Memorandum, Langley to Wallops, 14 September 1960, in folder "Wallops, September - December 60," in RGA181-1(C). Table, "Chronology of Flexible-Wing Research, NASA-LRC," 8 June 1971, in folder "Flexible-Wing," MA Collection. Shortal, 662-63; Hallion, 137-40.

94. The station at Woomera was moved to Carnarvon, and the Zanzibar station was evacuated after a revolution on the island. *Data Book II,* 548, 593-94. See also: "Gemini Trackers Training At Wallops Facility," *Goddard News,* 18 November 1963, 6.

95. Teletype, NASA GSFC Offices to ADE/NASA et al.," 21 November 1963, 2, in folder "Special File, May 63 - February 64," in RGA181-1(S). Memorandum, E. J. Stockwell for the Record, 15 November 1963; Letter. Edmund C. Buckley to General Antonio Perez-Marin, 14 December 1962, both in "Chron. File, January - December 62," in NASA HQ box #1.

96. *Data Book II,* 547. NASA's third network, the Deep Space Network, was operated from the Jet Propulsion Laboratory. In 1972 the satellite network and the manned space flight network were merged at Goddard.

97. U.S., Congress, House, Committee on Science and Astronautics, *Authorizing Appropriations to NASA, House Report 1240 to accompany H.R. 10456,* 88th Cong. 2nd sess., 6403-18H, p. 66.

98. U.S., Congress, House, Committee on Appropriations, *Independent Offices Appropriations for 1965, Hearings before a subcommittee of the Committee on Appropriations,* 88th Cong. 2nd sess., 6404-07H, 1239.

99. Teletype, W G Burton to BDA/STADIR M&O, et al., 11 July 1963, in folder "Wallops, January - June 46 [sic]," in RGA181-1(C).

100. U.S., Congress, House, Committee on Science and Astronautics, NASA *Authorization for 1965, Hearings before the House Committee on Science and Astronautics on H.R. 9641,* 88th Cong. 2nd sess., 6402-18H, part 4, 2129.

101. For Saturn I - Pegasus see: U.S., Congress, House, Committee on Science and Astronautics, *NASA Authorization for 1966, Hearings before the House Committee on Science and Astronautics on H.R. 3730,* 89th Cong. 1st sess., 6502-17H, 203; 255 for Scout. The Saturn I launches also supported Apollo testing, Data Book II, 58. For Saturn V see: "Schedule of Firings," 10 March 1965, in folder "Wallops, January - June 46 [sic]," in RGA181-1(C).

102. *Langley Researcher,* 31 May 1968, 8; Table "Chronology of Flexible-Wing Research, NASA-LRC," 8 June 1971, in folder "Flexible Wing," in MA Collection.

103. "Milliner," OHI, Tape 1b: 555.

Chapter 4

SPACE SCIENCE RESEARCH

The nature of the service provided by Wallops Station shifted during the transition era as the scope of research broadened and the customer base grew. Under the NACA, Wallops existed to serve the program needs of Langley, and occasionally Lewis, Labs. The research examined a range of aeronautical engineering issues, but whether delving into fundamental principles or focusing on a particular airframe, the program came through Langley. NACA Headquarters consisted primarily of administrative staff, not the centralized program offices that characterize NASA. Due to this management structure, Wallops really had to answer only to Langley. The military was a regular customer at Wallops, but again, the projects came through the "Mother Lab." The establishment of NASA, and the growth of the Space Race, provided stimulus for the expansion of the base. These factors also brought the new programs and clients that allowed the Wallops personnel to expand their niche within the organization.

As transonic and supersonic wind tunnels became more effective, the focus of military testing shifted. Hypersonic and high-temperature research for all areas of DOD soon replaced the launching of model airframes.[1] Most of these tests were conducted in a generally open manner, as such things go. The raw data held little value for anyone prior to reduction (the process of converting data readings into useable form), and truly sensitive hardware rarely appeared at the base, so the need for oppressive security measures seldom arose.[2] Many of the tests involved small pieces of larger puzzles, providing incremental progress to the researchers in a relatively short period of time. Occasionally, longer term projects like Trailblazer or RAM occupied the Station's attention, but since even the Scout had weight limitations compared to boosters available at the Cape and Vandenberg, small projects were the norm.[3]

One somewhat ironic exception to this norm resulted in both a Thor and a Jupiter missile coming to Wallops for tests in 1963. While still prohibited from developing the capability to fire the liquid-fueled boosters, the engineers erected the vehicles, filled them with water to simulate fuel, and began measuring how the winds at ground level stressed the airframes. The goal was to provide full-scale data to validate wind tunnel readings. Wallops could perform the tests with meteorological equipment already in place, without the need of tying-up an active launch pad at the Cape.[4]

Experiments also continued to come to the Station from within the parent organization. Langley remained a primary customer, and Lewis also continued to utilize the range. In addition, there were new members of the team to accommodate. Goddard became as large a source of projects as Langley, and even the Marshall Center, occasionally came to Wallops.[5] As with the military projects, NACA/NASA projects shifted from being almost exclusively aeronautical, to a mix of hypersonic and space science research. As time passed the shift became more pronounced, especially as three new types of customers began to arrive at the base: universities, non-military government agencies, and researchers from other countries.

It is important to realize that the term "space science," refers to more than just astronomical research focusing on celestial bodies. According to one NASA definition the term "space science" included, "theoretical and experimental research on the ground and in the earth's atmosphere," ..., [and] "also includes instrumentation development and directly-related supporting research and technology required for carrying out [these] investigations."[6] The wide range of topics that could be included under this broad definition attracted many research proposals. The NASA leadership, concerned with both facilitating basic research, and demonstrating practical applications of these publicly funded activities to Congress, developed a detailed organizational structure to coordinate space science research.

Under the leadership of Homer E. Newell, Jr., first as an Assistant Director of Silverstein's Office of Space Flight Development and later as an Associate Administrator in charge of the Office of Space Science and Applications (OSSA), a series of program offices was established at Headquarters to pursue research in differing fields.[7] A Space Science Steering Committee with associated subcommittees, reminiscent of the old NACA organizational style, worked to bring NASA personnel and researchers from outside the organization together to evaluate research proposals, and set priorities. The Space Sciences Board of the National Academy of Sciences played a similar role from a position outside of NASA.[8] Field centers had a voice in this planning function, but projects required OSSA approval in order to proceed, especially given that the November 1963 NASA reorganization placed the space science centers (Goddard, JPL, and Wallops) under Newell's cognizance.

From Wallops' stance this represented only a minor change from past practices. NACA projects came to the Station through PARD. For NASA, though field centers could "initiate proposals for scientific investigations and flight projects," approval from above was still mandatory, just from a different location.[9] "Since endorsement of the scientific objectives of any proposal must be obtained by the appropriate Headquarters Program Director, you should refer requesting organizations to deal directly with the cognizant program offices in Headquarters. This procedure is necessary to ensure the evaluation of a proposal not only in terms of the proposal's scientific validity but also in terms of its compatibility with on-going and planned NASA program efforts. ... Final approval for the implementation of any [research] program will be by the Administrator. In no case will any commitment of NASA resources or facilities take place prior to such approval.[10] Wallops' background as a service center lent an air of normalcy to this type of relationship with higher authorities, perhaps helping the Station to avoid the type of frictions that arose between Goddard and JPL, and OSSA.[11]

An October 1960 planning document, "Long Range Thinking in Space Sciences," listed nine areas of scientific investigation to be pursued from Wallops. These areas included meteorology, atmospheric motions, and solar studies.[12] Flights investigating atmospheric electron density and atmospheric probes joined the heat-transfer and materials research on the island's flight-line. Add the Mercury and Scout work to the roster and the result is a very busy schedule. While early research, like Shotput which supported William O'Sullivan's Project Echo, came from within NACA/NASA, an increasing number of universities began to avail themselves of the emerging scientific sounding rocket program at Wallops.[13] Just as World War II opened the way for physicists to receive major government funding for their research, Sputnik and the Cold War resulted in increased government funding for space science.[14] Unlike the investigations of the past, conducted mostly with passive devices like telescopes, the development of the sounding rocket allowed instruments to be lifted to altitudes unreachable by air-breathing vehicles and balloons. Also, while satellites promised to provide a powerful new tool for researchers, orbital altitudes could not drop much below 100 miles or the atmosphere would drag the vehicle down. Sounding rockets, often based on military designs, provided access to this intermediate zone. Since the government controlled most of the affordable hardware, and requisite facilities, universities looked to Washington for both logistic and financial support.[15]

The research projects conducted by the Universities of Michigan and Maryland in 1958 paved the way for other universities to conduct experiments at Wallops. These projects resembled the earlier NACA projects in variety, if not in subject matter. Some concerned "pure," fundamental research while others pertained to applied research of interest to both scientists and the military.[16] The military, of course, provided financial support for many programs pertaining to their needs. Wallops provided a few research grants, however a large portion of the basic research grants came from NASA Headquarters, through OSSA.[17] The research team "would produce their own hardware, and they'd interface with us here at Wallops, ..., so it would fit into the rockets, and we'd figure how high they wanted to go and what they wanted to do, and guide them so they could get the experiments [accomplished] ...."[18] The size and civilian orientation of the facility allowed professors to launch small projects and train graduate students without interfering with operations at the large ranges. This educational aspect of Wallops' operation helped not only to advance the cause of science, but it also provided a supportive clientele that would "rise up in arms" in response to later attempts to close the base.[19]

Another new group of customers at the Wallops base during the early 1960's consisted of government agencies outside the Defense Department and NASA. These organizations began to apply space age technology to their missions and found sounding rockets to be a valuable tool for certain tasks. In May 1963 Langley collaborated with the Atomic Energy Commission on a payload launched by Scout from Wallops. The Re-entry Flight Demonstration Experiment - 1, provided data relating to NASA-AEC efforts to build reactors for spacecraft that, in the event of an emergency, would be safely destroyed upon re-entry into the atmosphere.[20] While this "model" reactor carried little radioactive material, Krieger saw the need to formalize procedures for dealing with such substances. "In the past, this station has had negligible opportunities for support of research efforts in which radioactive sources were involved; however, in anticipation of increased contact with radioactive material, definite radiological safety regulations, procedures, and requirements applicable to all radioactive sources and experiments will be established ...."[21]

In June 1964 a researcher with the National Bureau of Standards proposed cooperation with NASA on Project RAM, through use of equipment the NBS already had in place at Wallops, "in support of most of the rocket-probe experiments conducted there."[22] The runways at the Main Base were utilized in cooperative projects with the FAA, which included research on tires, and landing aids.[23]

The Weather Bureau became a frequent, then a permanent, customer at the base. Even though Project Hugo proved impractical, the Bureau's hurricane research benefitted from work at Wallops. In August 1958 the FPS-16 radar at the base tracked Hurricane Daisy as the storm moved along the coast.[24] NASA planners conceived a two-front approach to meteorological research. The first involved building on past programs with balloon launches and sounding rockets to probe the atmosphere. Both systems offered advantages. Balloons could remain aloft for a long period of time, while sounding rockets could reach to the fringes of space. A combination of the two systems resulted in sounding rockets that could release a balloon at a high altitude, allowing data to be taken during the balloon's long, slow descent. The establishment of a "meteorological network," gave atmospheric researchers the opportunity to obtain data from several areas around the world.[25]

While participation in this network acquainted Wallops with a novel type of rocket, and tied into Langley's investigation of wind shear phenomena, Krieger was not overly enthusiastic about the quantity of launches the weather program would entail. "Mr. Krieger says that participation in the Meteorological Network does not provide material assistance to his operations and therefore, as far as Wallops is concerned, it has no interest in the network once the present program is completed. However, if some group within NASA felt that further participation was justified for meteorological research he would be willing to continue although it does represent a fairly heavy workload for Wallops."[26]

These misgivings did not fade easily. Seventeen months later Krieger reported, "It should be pointed out from a range's viewpoint that the present operational status of meteorological rockets requires considerable range time; and launching, tracking and data facilities in addition to range clearance and surveillance of impact areas. Therefore, much range effort is required with the present state of the art, and should be considered in the selection of rockets, payloads, number of firings and scheduling."[27] A month later he added, "Being on the operational end of this business, we have been somewhat disappointed for it seems to us that this program turned from a research and development effort into an operational effort almost immediately after its initiation."[28]

Despite the early problems Wallops, and Krieger, adjusted to the meteorological program. In March 1963 developmental launches of a Nike-Cajun meteorological rocket began, and by April 1965 "a regular launch schedule [had] been established," for the smaller rockets that performed most of the weather flights.[29] By late 1965, perhaps because of the usefulness of the program for both science and justification of budgets, Krieger accepted that the meteorological research occupied a prominent place in the Station's agenda.[30]

The second portion of NASA's weather program involved the construction and launch of an orbiting satellite. Inherited from the Army Ballistic Missile Agency when NASA was established, a meteorological satellite program contracted to RCA came under Goddard's jurisdiction. The resulting TIROS (Television Infra-Red Observation Satellite) revolutionized weather research and forecasting, and provided a high-profile example of an everyday application of space technology. After successful launches of the first two TIROS, plans were made to move the data reception station from the Army base at Fort Monmouth, New Jersey, to Wallops Island.[31] RCA already held the contract for operation and maintenance of Wallops' FPS-16 radar, and could thus easily expand on its existing activities there.[32] NASA hoped to relocate the data reception equipment in time for the launch of *Tiros* 3 in July 1961, but the work went faster than anticipated and the facility became operational a month early, in time to support *Tiros* 2, already on orbit. Though most of the funding came from the Weather Bureau, Edmund Buckley requested funds to cover the extra personnel costs incurred in running the radar and processing the received photographs. In December the facility underwent an overhaul.[33]

By late 1962, in keeping with its research and development mission, NASA was planning a second generation weather satellite called Nimbus. The Weather Bureau, pleased with Tiros however, began planning the "Tiros

Operational System" (TOS), and in 1964 pulled out of the expensive Nimbus effort.[34] Wallops saw both sides of the controversy as NASA considered the base for a backup Nimbus station, while simultaneously upgrading the Tiros facility.[35] Whatever difficulties the NASA-Weather Bureau spat caused at higher levels of the organizations, working relationships at Wallops seem unimpaired. In September 1964 NASA recommended that the East Coast Command and Data Acquisition station for TOS be located at the base. The presence of this operational system required some balancing of projects at the test range, but "The Director of Wallops has indicated that he will cooperate to the fullest in order to prevent conflicts between [different projects'] operations." By this time, the meteorological sounding rocket program had provided Wallops with experience in the conduct of an operational program supporting Weather Bureau needs, so this work integrated into the Wallops program without too much trouble.[36]

The increase in Wallops' obligations and range users brought about a new round of construction funding. The initial post-Sputnik infusion of money into the base, $20 million for Project 2080 in fiscal 1959, was followed in fiscal 1960 by $0 in the "Construction of Facilities" account. NASA's request for fiscal 1961 contained a modest $2.03 million in Wallops CoF for a computing station and a vehicle check-out building.[37] The $13 million request in fiscal 1962 represented a hefty boost in the Station's budget. $1 million of this money went to complete repairs to damage caused by the Ash Wednesday Storm, the remaining amount bought improved tracking equipment and better vehicle handling facilities.[38]

In addition to more fixtures the Station also doubled the size of its workforce within three years, growing from 229 paid employees in mid-1960 to 493 by mid-1963.[39] The number of contract personnel grew in roughly equal fashion. "When we became NASA we realized that, some of us, we were not going to get as many people as we thought we needed. ... The magic number has always been 1000 to 1100 employees. When we would cut back the civil servants we seemed to be able to fill in with contractors.[40] While these personnel and funding increases might have drawn serious Congressional scrutiny in other years, during the early stage of the Project Apollo expansion, it constituted a drop in the bucket. The construction budget for Cape Canaveral in fiscal 1962 topped $115 million. The same three years in which Wallops doubled its complement saw the in-house employment at the Manned Spacecraft Center rise from 668 to 3345.[41] Granted these are extreme examples, but they illustrate the point; the growth at Wallops was in the "enviable position" of being overshadowed by other, more visible growth in the space program.[42]

Coordination of the rising launch rate with the incorporation of new facilities into the Wallops infrastructure called for efficiency in the Station's operations. As early as April 1960 Abe Silverstein felt compelled to issue a memorandum that addressed the subject of, "Launching Schedule Problems at Wallops Station." Directed to the two primary range users, Langley and Goddard, the memo emphasized the need to maintain firm flight schedules. "Continual changes to the established monthly schedules is becoming a source of embarrassment to Wallops and NASA Headquarters ...."[43] To help alleviate these problems, and refine the processing flow of projects, the Wallops staff issued a four volume "Wallops Station Handbook," in December 1961. The cover letter that accompanied the first edition stated, "This 1961 edition of the WSH is the station's first attempt to outline the policies, procedures, facilities, and organization of the station for the guidance of prospective range users." These policies and procedures went promptly into effect.[44]

One potential extension of Wallops' capabilities that NASA decided not to authorize resulted in limiting the growth of the Scout program on the island. The question of launching into high-inclination or polar orbits from Wallops had been considered, and in April of 1961 two studies recommended against the proposal. The primary reasons involved range safety issues. Such launches would come too close to inhabited areas, so Scout payloads requiring these orbits continued to fly from Vandenberg Air Force Base.[45]

Coordination with the military ranges did not always prove easy, as the civilian range often found itself overlooked.[46] In May 1962 Buckley sought to correct this situation by proposing, "that the appropriate agreements and policies be established which would allow the NASA Wallops Station to be included in [the] lead range concept." This concept served to facilitate operations by giving the designated "lead range" the assignment of planning, coordinating, and controlling the requisite facilities for a particular launch. Acceptance of this proposal allowed Wallops easier access to components of the Atlantic Missile Range (and vice-versa), and also gave the Station more credibility as an "integrated range," rather than as simply a "launch site."[47]

The designation of Wallops as a lead range occurred as the Scout launch facility completed a refurbishment in July 1962. This refurbished pad still did not provide the base enough capacity. "A two per month Scout launch capability is required by the summer of 1963 to meet the NASA launch schedule and additional Department of Defense missions."[48] To meet this need NASA took advantage of its ability to reprogram certain appropriated funds and sidestepped Subcommittee Chairman Albert Thomas by allocating $1.5 million to begin construction of a second Scout pad.[49] Even though its program financially paled in comparison to larger space initiatives, Wallops did not always get what it asked for. The Senate removed $2 million from the fiscal 1964 budget, having deemed a planned sounding rocket launch facility (intended as Wallops' #6 launch pad), "desirable, but not essential at this time."[50] The previous year both House and Senate Authorization Committees had imposed a 5% cut in funding at several field centers, including Wallops.[51] Still, with the pads built during Project 2080, the Scout pads, and an Aerobee launching tower that had commenced service in 1960, the total launch capacity at the Station was now varied and ample.[52]

The tracking and data acquisition program at Wallops diversified as the variety of experiments increased. In June 1961, while Wallops oversaw the operation of the Mercury tracking station in Bermuda, Langley requested permission to build and operate, "a temporary base for ... mobile tracking trailers," at Coquina Beach on North Carolina's Outer Banks. This station would provide down-range support for longer flights such as RAM. Abe Silverstein agreed to the proposal, but stipulated that, "when certain range instrumentation now under development at Langley becomes operational, and is assigned to Wallops for integration with other equipment ... as a part of a *continuing* requirement in support of Wallops launchings, it appears logical and necessary that Wallops assume full responsibility, with, of course, continued co-use by Langley as required."[53] In May 1965 contracts relating to the North Carolina facility were transferred from Langley to Wallops.[54]

Tracking operations continued to migrate off the Wallops base when, in May 1962, Wallops contracted with the Military Sea Transport Service (MSTS) for the operation and maintenance of a tracking and recovery ship, USNS *Range Recoverer*. Acquired and outfitted with assistance from Langley, Wallops personnel handled the tracking equipment while leaving the nautical tasks to an MSTS crew. As Buckley told Congress, "we do feel it's necessary for us to operate the instrumentation on a ship, so that the whole ground system works together as an integrated system."[55] Equipment aboard included helix antennas, recording systems for data, and a communications suite with direction finders to help locate and recover payloads. Wallops had used shipboard tracking systems before, calling on the Navy for assistance and occasionally renting ships. The accelerated launch schedule called for a level of support that made it practical for Wallops to have a ship of its own.[56]

The increased range of the vehicles launched from Wallops also meant that a greater area of ocean had to be cleared of shipping. "Range safety regulations dictate that the probability of impacting a ship shall be less than 1 in 100,000." An SPS-12 radar at the base could search areas out to around thirty miles. A shot travelling much further than that required personnel to go out and verify that the range was clear. The *Range Recoverer* helped with this task, but the ship's slow speed and the limited range of its surface search radar minimized its effectiveness in this role. To fill in the gaps in ocean surveillance Wallops relied on aircraft supplied by different organizations. Also useful for calibrating tracking systems, finding sources of radio-frequency interference, and locating downed payloads, surveillance aircraft became a vital tool in Wallops' arsenal. So much so, that in September 1964, Krieger requested that Headquarters transfer two Langley planes to Wallops, and acquire a Lockheed C-121 to meet the Station's needs. NASA met Krieger halfway, they arranged for a Lockheed contract for two C-121's to cover Wallops.[57]

The Navy provided much support to Wallops at this time. Aircraft for range surveillance, divers to recover test components, and ships for tracking purposes.[58] There was even a proposal to use Navy airships from Lakehurst, New Jersey, to recover missile nose cones. A nose cone and parachute were sent to Lakehurst in 1961 so airship crews could practice recovery techniques, but the plan fell through when the Navy retired its airship fleet later that year. Soon after that, Wallops contracted for the *Range Recoverer*.[59]

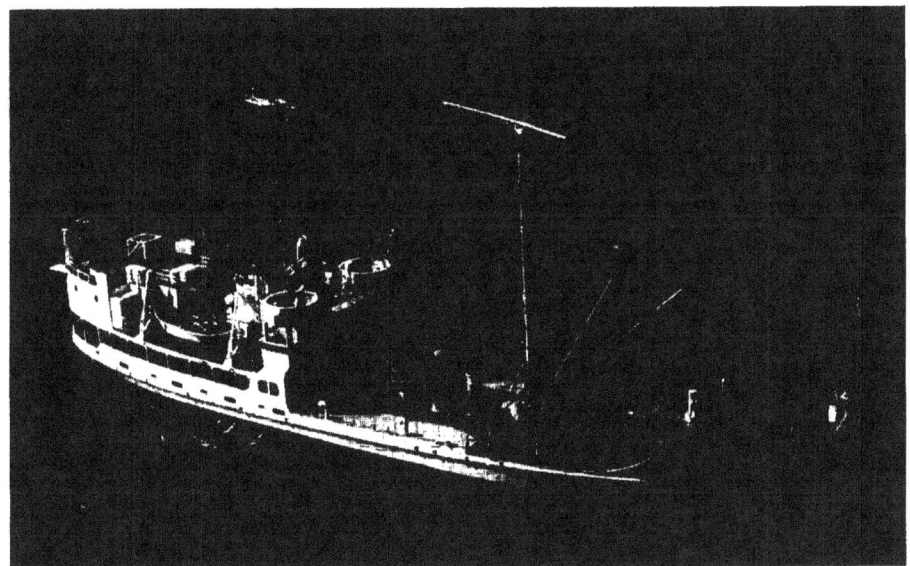

Aerial view of the Range Recoverer.

One facet of the space program new to many in NASA stemmed from section 102, paragraph c (7) of the Space Act. "The aeronautical and space activities of the United States shall be conducted so as to contribute materially to ...: Cooperation by the United States with other nations and groups of nations in work done pursuant to this Act and in the peaceful application of the results thereof; ...."[60] Since a high-level concern revolved around the possibility that some nations might react to Sputnik by strengthening their ties with the Soviet Union, both Executive and Legislative decision makers wanted NASA to include foreign governments in selected portions of the space program. This would provide the opportunity to showcase U.S. space technological prowess, highlight Soviet secretiveness, and permit other nations to advance their scientific and political stature.

A study of the political motivations of NASA's international program characterized official sentiment toward this issue as being either "innovative," or "conservative."[61] Innovators believed that cooperation could lead to more peaceful international relations, an opening of the Iron Curtain, and the creation of a new political reality. Conservatives saw the opportunity to wield a potent propaganda tool, and stimulate research and the creation of scientists, engineers, and technologies needed for more effective maneuvering within the existing reality. Cooperative projects were therefore useful to both ideological camps, only for different reasons. An examination of four areas of cooperation concluded that although NASA, on the whole, spoke the language of cooperation like an "innovator," the agency followed a "conservative" path out of political necessity.[62]

No matter how well this conclusion applied to NASA as a whole, it does not adeqately describe Wallops. While "conservative" motivations, particularly propaganda value, were recognized at the Station and Wallops' major role in the program provided NASA with ammunition to fight later proposals to close the base, the value of the science performed and a perception of being NASA ambassadors seemed to have also been consistent "innovative" motivations.[63] The explanation for why this should be the case, and why Wallops received a large share of the program, stemmed from several roots.

The NACA had almost as little experience with cooperative foreign projects as it had with public relations. Established specifically to advance the state of aeronautical research in the U.S., in response to European progress, the Committee usually saw competition when it looked abroad. Also, the NACA did not hold the near monopoly on aeronautical technology that NASA later held on space technology. NACA officials traveled to Europe to learn and observe rather than to teach and the NACA era was almost at an end before any non-American researchers came to Wallops.[64]

Through their experience with the International Geophysical Year, many of the Project Vanguard researchers who constituted the core of Goddard had at least some grounding in the organization and conduct of international cooperative ventures. The success of the IGY's wide ranging research program of atmospheric and geophysical studies, performed under the auspices of the International Council of Scientific Unions, demonstrated that such ventures could produce good scientific results. Scientists like Homer Newell and Milton Rosen had both influential positions within NASA and recognition in the international scientific community. The growing role of the research techniques in use at Wallops, relatively inexpensive rockets familiar to these scientists, made the Station a logical place from which to launch cooperative projects.[65]

One NACA veteran who had gained some experience in international relations, and also occupied an influential position within NASA, was Edmund Buckley. His long association with Wallops provided the base with a friend at Headquarters who would look-out for its interests. The decision to locate the Mercury demonstration tracking facility at Wallops, for the convenience of both the Space Task Group, and Tracking and Ground

Instrumentation Unit at Langley, resulted in many foreign nationals visiting the area during the course of this program. The negotiations Buckley participated in no doubt familiarized several governments with Wallops' overall potential for science research.[66]

NASA Headquarters established the Office of International Programs on 1 April 1959 and assigned it the task of coordinating NASA's cooperative ventures. This included assisting Buckley in locating sites for Mercury stations, and coordinating the international use of the Wallops range.[67] In December they reported to the Administrator an agreement reached with Canada for cooperative launches from Wallops. Since the Canadians had been the first non-US group to use the base, this hardly represented a political breakthrough.[68] As the differing components of NASA settled into their roles, however, the program made more progress. Though an earlier House committee report on the subject of "International Cooperation in the Exploration of Space," made no reference to Wallops, by May 1960 Hugh Dryden told Congress that, "This station will be the scene of launchings made in cooperation with other nations in our program of international cooperation in space activities."[69]

From 28 to 30 June 1960 the "second working group meeting," for the International Ionospheric Satellite (a joint US-UK project) was held at Wallops. The resulting *Ariel 1* satellite represented NASA's first joint satellite project and was launched from the Cape in 1962. Two of four subsequent Ariel missions launched from Wallops.[70] In January 1961 a French professor and a Norwegian engineer both arrived at the Station to work on separate tracking systems. In March, France announced plans to launch soundings from Wallops, and a joint US-Italian agreement was formalized. Over the course of the next several years, researchers and trainees from Australia, Japan, Sweden, India, and Pakistan came to the Station in association with a variety of projects. (See appendix 4).[71]

The introduction of people from different cultures and backgrounds into the local Virginia communities caused some frictions at first, as might be expected. "We had a few problems, like getting haircuts and motels, but we worked it out behind the scenes, and most people were pretty decent about it. This is back in the time before integration, you know, and these people, a lot of them, were dark-skinned, but they spoke beautiful English, most of them, ..., each country sent its best."[72] As time passed and people came to know some of these visitors, the problems eased. Indeed, during the three years the Italian team was here, "some were married to some of the local girls around here and they had started families." Krieger noted that the Italians, "really got to be a part of the family here. We almost had tears in our eyes when they left."[73]

One method the staff used to educate the public about the programs, foreign and domestic, in progress at the base was a series of annual "Open House" events. The first of these events took place in October 1962, included self-guided tours, literature describing the Station and its programs, and the opportunity to talk to NASA employees and witness rocket launchings. Besides helping to strengthen ties between the base and the community, the open houses allowed Wallops to showcase the open, civilian nature of its work.[74] The ability to demonstrate such accessibility without compromising either national security or the launch schedule of a multi-million dollar project provided one of the major justifications for bringing the international guests to Wallops. The issue of appearances counted for much, especially as the US endeavored to score propaganda points against the USSR. Joe Robbins stated, "We had to be very careful, that when foreign people came here, ..., if we had any military, I instructed my people to go around and tell every one of them, 'Don't come on this base wearing a uniform; if you do you're not going to be admitted. You're going to be dressed in civilian clothes.' As a matter of fact, I had to tell a general that too. ... What would it look like if you were bringing a foreigner in and you had half the people you see on the base were military? Going back to their own country, 'Well, they just say it's civilian, but it's all military.' ... We tried to keep a clean slate."[75]

Specific cooperative research projects covered a wide range of services. Projects that utilized flight hardware included launches of foreign payloads, coordinated launches from Wallops and other ranges, and the study of rocket preparation and launch techniques. Researchers also studied and received training on the assorted tracking and data acquisition equipment. Using surplus radars Wallops put together an "instrumentation lending library," where foreign customers "could borrow a radar, borrow a telemetry system, ..." Some visitors simply came to tour the Station and observe a portion of the American space program close-up.[76]

These projects permitted the participating countries to conduct space flight research and advance their technological base without having to expend the enormous, or (for some) impossible, sums of money required to build a massive space transportation infrastructure. Sounding rockets, balloons, and the Scout were comparatively cheap, and the operation of these vehicles entailed small, low-budget facilities similar to Wallops in scale if not in diversity. This should not be taken to indicate that the scientific data generated had little use. The establishment of sounding programs at several points around the globe gave researchers the ability to study atmospheric conditions at a variety of locations, and helped create a comprehensive picture of the Earth's climatic system. NASA's concentration on high-priority missions like Apollo, Tiros, and the Orbiting Observatories, prevented the agency from establishing a large number of small facilities to gather this data. By cooperating with interested governments, they could at least obtain access to the data coming from these smaller programs.

The most prominent example of this mutually beneficial international cooperation was the San Marco Project conducted with Italy. After a series of sounding rocket launches from both Wallops and Sardinia, the Italian

Space Research Commission expressed a desire to utilize the Scout booster for satellite research. Arrangements were completed by April 1962, and the US-Italian team went to work.[77] On 15 December 1964 a group of Italian technicians launched *San Marco 1* from Wallops, "the first satellite entirely designed and constructed in Europe," on a mission to analyze the fringes of the atmosphere.[78]

Not content with launching only from Wallops, the Italians commenced construction of a mobile launch facility in Formosa Bay, off the East Coast of Africa to provide a Scout launcher near the equator. The San Marco Equatorial Mobile Range contained two platforms anchored to the sea-floor, and an on-shore base in Kenya. One platform provided for assembly, testing, and launch of the Scout, the other served as the launch control facility. A tracking station, integrated into NASA's STADAN network, supported operations from the Kenyan base. Owned by the University of Rome, and operated by the Italian Aerospace Research Center, the facility was completed in 1967 and launched *San Marco 2* into orbit on 26 April of that year. On 12 December 1970 the range launched *Explorer 42*, the first satellite launched for the U.S. by another government.[79]

In 1965 Wallops began to coordinate its meteorological soundings with those of Argentina and Brazil to create the Experimental Inter-American Meteorological Rocket Network (EXAMETNET). With OSSA at Headquarters overseeing the project, and Langley providing the hardware, Wallops participation in this project enabled, "comparative analysis of the structure and behavior of the atmosphere in both hemispheres."[80]

Not only did foreign researchers come to Wallops, but Station personnel traveled abroad to assist in establishing ranges, and launching rockets. In addition to helping the Italians with both sodium vapor launches from Sardinia, and the San Marco Project, Wallops people travelled to Sonmiani Beach, Pakistan, to help set up that launch site.[81] Cooperation with Canada included launches by Wallops personnel from the Ft. Churchill range in Manitoba. In a cost sharing arrangement with the U. S. Air Force, which also used the range, Wallops maintained facilities at that launch site. After a fire heavily damaged the facility, Canadian engineers brought their Black Brant test program to Wallops for a time, while the US participated in the restoration of Ft. Churchill.[82] As a result of the international programs, "Wallops is better known overseas, in a way, than it is here in this country, ..."[83]

Intimate knowledge of Scout procedures took Wallops engineers to other American launch ranges to provide assistance. A meeting held at Wallops on 14 April 1961 with representatives from Headquarters, Langley, Marshall, and Vought Corporation, discussed the requirements of the Scout facility at the Point Arguello (Pacific Missile Range) launch site. Marshall's involvement did not last long, as in August the decision to transfer "technical cognizance" of the Scout from Langley to the Huntsville facility was reversed. Werner von Braun, working on the Saturn booster for the recently accelerated Apollo Project, probably lost little sleep over this decision.[84] Wallops dispatched personnel to the Pacific Range several times over the course of the next few years to help with problems like "pyrotechnic procedures," or to "provide technical consultation on the dynamic balance" of various Scout missions.[85]

In mid-1964 Wallops engineers participated in the investigation into the cause of a fatal accident at Cape Canaveral. While attaching an Orbiting Solar Observatory to an X-248 rocket motor (the third stage of the Delta booster assigned to loft it into orbit) static electricity activated the motor's ignitor. The solid-fueled motor fired inside the test building, injuring eleven technicians, three of whom later succumbed to their wounds. Since the X-248 also served as the fourth stage of the Scout, Goddard (responsible for the Delta) included representatives from both Wallops and Langley in the fact finding board.[86] Those assigned to the board lent their experience with the X-248 to the investigation, and learned all they could to prevent a similar accident from happening at Wallops.[87]

The operation at Wallops was not accident free during this period, though fortunately no-one died. On 8 June 1964, a meter measuring the flow of nitrogen gas in a test apparatus exploded under pressure, injuring the Langley engineer running the test. While directly caused by the improper installation of a valve, the accident report cited a number of contributing factors that added up to trying to do too much, too fast.[88] The acceleration of the space program, and especially the influx of experimenters inexperienced with the rocket operations, made "keeping track of everything that was going on ... sometimes pretty difficult." "[Krieger's] philosophy was you don't do anything unsafe, ..., even if you're handling rockets you never do anything unsafe because they're not safe devices; ..., and assume that you're going to screw-up, because humans do make mistakes."[89] This philosophy, combined with the experience of a rocket misfire in 1951 that had cost a technician his right hand, created a safety consciousness that spread throughout the Station.[90] In 1962 Goddard and the University of Michigan brought the ARGO D-8 radio noise probe to Wallops for launch. "A launch was set for the last week in July, but the Range Safety at Wallops postponed the launch to an unspecified future date so that they can evaluate the ARGO D-8 performance limits as related to a Wallops launch." ARGO D-8 flew on 22 September.[91]

The off-range experience served the Wallops engineers well when, in May 1964, they began to conduct Nike-Cajun firings from a new launch site at Point Barrow, Alaska. Situated near the Arctic Research Laboratory, the new site gave NASA a launch range in the far north, especially beneficial to the meteorological sounding program. A survey team from Headquarters, Goddard, and Wallops considered several sites and decided that, "Point Barrow was by far the most favored by reason of its desirable location, and the presence of an active research laboratory housing upwards of 350 scientists and technicians, a DEW line station, an NBS ionosphere sounding station, a W.B. station and adequate commercial air service."[92] Working out of Quonset huts, and restricted to setting up equipment only during the brief Arctic summer, the Point Barrow program was reminiscent of the early years at Wallops. While OSSA and Goddard handled the program management, OTDA and Wallops constructed the NASA facility. Wallops also had the responsibility for, "preparation and launch of vehicle," and, "training of launch crew."[93] Support and year-round maintenance was furnished by the Arctic Research Lab "on a cost-reimbursable basis." Though the operations "up north" put "a strain on the travel budget," it did not cause exceptional administrative headaches for the Wallops staff; "just a logistic problem of getting things there, we were used to doing [off-range launches] anyhow, ..., it didn't make that much difference."[94]

By far the most ambitious off-range operation conducted by Wallops during this era was the Mobile Sea Launch Expedition of 1965. From the beginning of serious U.S. space planning the concept of launching rockets from ships at sea attracted attention.[95] Launching from sea complicated missile navigational planning, but it would provide mobility, thus allowing the launch platform to be placed in an advantageous spot. The Navy made use of this advantage by launching sounding rockets from a ship positioned in the path of a solar eclipse in 1958.[96] The open expanses of water would also provide a margin of safety, especially for the larger or nuclear powered boosters then under consideration. In 1961, the House Committee on Science and Astronautics convened hearings on the concept as it related to the possible establishment of an American launch range along the equator. During these hearings, NASA depicted the use of ships to launch large boosters as uneconomical and unnecessary, but, "a continuing requirement exists for shipboard launch of sounding rockets in the equatorial regions and the extreme southern latitudes."[97] No direct connection to Wallops appears to have been made during these proceedings, however.

The Office of Space Flight Programs made a request in August 1961 to use the tracking ship *Coastal Sentry Quebec,* then on station in the Indian Ocean supporting Project Mercury, in conjunction with the sounding rocket project underway with Pakistan. The desire was to utilize the ship's equipment for tracking launches, and for the ship itself to conduct rawinsonde operations.[98] Wallops' association with this project, through its assistance in training personnel from both Pakistan and India, meshed with their acquisition of the *Range Recoverer.* "The whole space science thing generated a lot of interest, and as young people we jumped on the bandwagon, and thought up projects."[99]

By November of 1962, the engineers at Wallops had formulated a plan to lease a small aircraft carrier from the MSTS, and outfit it with "roll-on instrumentation and bolt-down rocket launchers." "It will also be used to provide downrange telemetry tracking of rockets launched from Wallops Island, a platform to launch small two-stage rockets for obtaining meteorological data in the upper atmosphere, and a take-off and landing area for helicopters to be used for recovery operations."[100] The rockets utilized would be reliable Nike-Cajuns, and others of that, or smaller, size. The cost of contracts to prepare the ship and equipment was estimated at $191,050, and the operational cost placed at $3,700 per day.[101]

Unlike the other programs which came to Wallops, this one originated there. While Langley, Goddard, and other segments of NASA supported the plan, "this was one of the first jobs we went out and sold. ..., and they [Headquarters and Goddard] bought it, they bought it solid."[102] One of the reasons the project "sold" was the impending multi-national research effort for the International Year of the Quiet Sun (IQSY), which provided the ideal opportunity to use the mobile range concept. One of the motivating factors in the organization of the IGY was the occurrence of a period astronomers refer to as solar maximum in 1957 and 1958. Variations in the number of spots visible on the sun, and thus the amount of solar magnetic activity, exhibit an eleven year cycle. A period of many sunspots during the IGY permitted researchers to obtain information about the atmosphere and near-Earth environment, then repeat their experiments during the following period of minimum solar activity, and compare the two sets of data. The international scientific community organized the IQSY to coordinate the research activities associated with this 1964 to 1965 solar event.[103]

On 21 November 1963, OSSA's Director of Physics and Astronomy Programs, John Naugle, wrote to Stanley Ruttenberg, Executive Secretary of the US-IQSY Committee, informing him that, "NASA is considering a mobile balloon and sounding rocket expedition, to be carried out during the second half of 1964, in the broad ocean area of the southern hemisphere," and asking for research proposals. The attached distribution list indicates that copies of this letter went out to over seventy researchers in NASA, various groups associated with the DOD, and numerous universities.[104] Positive responses proposed meteorological soundings, studies of the Earth's magnetic field, and a few flights to support classified DOD programs. Preparations for the expedition commenced.[105] The game plan for the expedition included outfit of the carrier at the Port of Baltimore, followed by a shakedown cruise to a position off Wallops Island, where launches from both ship and shore would provide research data as well as practice operating the mobile range.

The plan almost received a serious setback when the Military Sea Transport Service reported that the carrier intended for use had suffered "extensive damage," precluding an East Coast shakedown.[106] Quick action by the service, however, gave Krieger "reason to hope that a Card class carrier can be supplied for the shakedown cruise."[107] The Service supplied USNS *Croatan* for NASA's use, an escort carrier that had seen use during World War II. MSTS provided the crew for the ship, Wallops provided most of the instrumentation and engineers, and various researchers and their NASA sponsors provided the payloads.[108]

The November 1964, shakedown cruise of the ship demonstrated the efficacy of the mobile range concept and the ability of the *Croatan* to perform its new mission. Twenty-five Wallops personnel conducted a series of rocket launches during the two week cruise, flying payloads provided by Goddard and the Universities of Michigan and Illinois.[109] Coordinated launches of sodium vapor rockets from both ship and shore launch sites took place from 10-12 November, followed by tracking tests on the 17th, and a launch to investigate electron density in the ionosphere two days later. After removing equipment back to Wallops, *Croatan* returned to Baltimore to be prepared for its mission to the Pacific.[110]

In January a convoy of vehicles left Wallops bound for Baltimore where the equipment was reinstalled, and preparations finalized. *Croatan* departed on 15 February 1965, spent two days off Wallops Island calibrating equipment, then headed for the Panama Canal.[111] After transiting the Canal, the ship left Balboa, Panama, on 6 March bound for Lima, Peru, and launched a series of ten sounding rockets while on route.[112] While in port, more equipment and personnel came on board, and "we opened it up so the public could go aboard and see what was going on." Local officials and media also toured the ship. This followed NASA's policy of demonstrating openness, while generating support for space science research, and touting the ship's peaceful mission.[113] The experience of conducting the open house events and hosting international researchers at Wallops served the engineers well here, as they already had practiced this type of public relations affair.

The ship sailed from the Port of Callao two days later and began the principal scientific portion of its mission.. Experiments were launched over the following three weeks for Langley and Goddard; the Universities of Illinois, Michigan, and New Hampshire; the Naval Ordnance Test Station, Air Force Cambridge Research Lab, and Sandia Corporation. Meteorological work conducted by the Weather Bureau, in support of the mission as well as for research, took place on a daily basis, and the National Bureau of Standards launched ionospheric soundings.[114] The ship then returned to Callao, off-loaded the equipment and personnel who had boarded at that Port, and departed for another three weeks of launchings.[115] The last launch of the mission occurred on 16 April, and the expedition officially ended later that week when *Croatan* made port at Valparaiso, Chile, where the engineers again conducted an open house. The NASA personnel then disembarked to fly back to the U.S., while *Croatan* returned to Baltimore where the mobile launch equipment was removed. The expedition had launched over eighty payloads, and numerous weather balloons and small meteorological sounding rockets. Five of the major launches occurred north of the equator, one on it, and the remainder from southern latitudes.[116]

The ship returned to carrying cargo, but the mobile launch equipment became a feature of the Wallops program. Engineer Robert Duffy stated, "That really led to the campaigns, nowadays Wallops is doing a campaign a year someplace. Up to that time space science was done at existing launch ranges, ..., we really cut our eyeteeth on that."[117] While the aircraft carrier turned out to be a little too expensive for regular use, Wallops used the *Range Recoverer* for similar missions, if on a somewhat smaller scale. The Wallops tracking ship travelled to Greece in May of the following year to conduct coordinated research on a solar eclipse. While in port there, Engineer Abraham Spinak and the other Wallops personnel met the Greek royal family; and, while only a part of the American contingent present, they continued to fulfill their dual roles in NASA's international program, science, and public relations.[118]

The influx of new customers, and the expanding role of the space science program in NASA, thus served to alter the scope of Wallops' operations. While still participating in aeronautical engineering research, the Station's mission changed to one of support for science research. The staff also became familiar with coordinating the activities of diverse groups in the performance of its mission. By the end of 1965, Wallops had changed in many ways, yet in other ways remained very similar to the place it had been in 1957.

NOTES

1. Letter, Abe Silverstein to Dr. L. A. Del[eceso?], 7 August 1961, in "Chron. File, July - December 1961," in NASA HQ box #1, for example of an Army request for two sounding rocket and five balloon launches. For examples of work for the Navy at this time, see: Flight Plan, LRC E135 3111, 13 December 1960, in folder, "Wallops, August 62 - May 63;" Letter, Robert L. Krieger to Commander U. S. Navy Ordinance Lab, 24 March 1964, in folder "Wallops, November 63 - March 64," in RGA181-l(C). For examples of Air Force projects see: Letter, Trygve Blom to NASA, 27 January 1961, in Chron. File, January - June 1961," in NASA HQ box #1; and chapter 2, above, for work on the Blue Scout then underway.

2. "Spinak, et al.," OHI, Tape la: 140-60.

3. Project RAM (Radio Attenuation Measurement) investigated the radio "blackout" that occurs when a spacecraft re-enters the atmosphere. It involved seven launches from 1961 to 1968: *Data Book II*, 464-66. For Trailblazer see chapter 2, note 125, above.

4. Memorandum, Wilmer H. Reed III and Robert M. Henry to Associate Director, LaRC, 5 October 1962, in folder "Special File, October 62 April 63;" Memorandum, Jerome T. Foughner Jr. to Associate Director, LaRC, 27 May 1963; Memorandum, Jerome T. Foughner Jr. for Aeroelasticity Branch Files, 13 June 1963, both in folder "Special File, May 63 - February 64;" all in RGA181-l(S). Here we see another idea that died hard; there are several references throughout this period to field level hopes of outfitting Wallops for the large, liquid-fueled boosters. See, for example: Memorandum, Carl A. Sandahl for Associate Director, IO December 1962, in folder "Wallops, August 62 - February 63," in RGA181-l(C).

5. See "NASA APR, April 1961", section arp, pp. 1.5, 1.6, 6.1, 8.2, for examples of Langley projects. For the 0 Gravity fuel experiments in support of Centaur engine development see "NASA APR, February 1961," section arp, p. 1.6; "NASA APR, April 1961," section apr, p. 1.5; "NASA APR, June 1962", c, 1.2; "NASA APR, July 1962," c, 2.10; and "NASA APR, June 1963," c, 2.4. Letter, Robert L. Krieger to Gerhard Heller, 11 May 1961, in folder "Special File, April - August 61," in RGA181-l(S). U.S., Congress, House, Committee on Science and Astronautics, *NASA Authorization for 1962, hearings before the House Committee on Science and Astronautics on H.R.3238 and 6029*, 87th Cong. 1st sess., 6103-13H, part 1, 24; for Robert Seaman's testimony regarding projects underway at Goddard, which include quite a few (Echo, Tiros, sounding rockets) that involve Wallops.

6. NASA Management Instruction #7100.1, 29 April 1964, attachment A, 2, in box, "NASA HQ Organizations, O.S.S.A. (con't)," in NHO.

7. *Data Book II*, 197-202. Some of the fields included: Lunar and Planetary Programs, Bioscience Programs, Geophysics and Astronomy, and Meteorological Systems.

8. Levine, 167-68. See also: NASA Management Instruction cited in note 6 above.

9. NASA Management Instruction as cited in note 6, above. On attachment A of this instruction is stated, "For the purposes of *this Instruction*, the term field center includes: ... (11) Wallops Station," [emphasis in original].

10. Letter, Thomas F. Dixon to Robert L. Krieger, 31 January 1962, in folder 005065 "15.1 Sounding Rockets to 1963," in file tray "Sounding Rockets - General," in NHO.

11. Levine, 16, 167-72. "At Goddard the differences between Director Harry Goett and Headquarters officials became so serious that he was dismissed in July 1965." Levine also points out that the contractual relationship between NASA and JPL was a source of difficulties.

12. NASA Headquarters, "Long Range Thinking in Space Sciences," October 1960, in box "NASA HQ Organizations, O.S.S.A. (con't)." The areas either directly referring to Wallops, or to Wallops related programs are: Meteorology, Upper Atmosphere, Ionosphere, Atmospheric Motions, Magnetic Fields and Particles, Solar Studies, Stars, Interstellar Matter, Shipboard Launchings. No role for Wallops is forecast in Lunar and Planetary Studies.

13. Shortal, 685-95, for Shotput and Echo.

14. Nathan and Ida Reingold, ed.s, *Science in America: A Documentary History, 1900-1939* (Chicago: Univ. of Chicago Press, 1981), 1-6.

15. *Data Book II*, 270-72.

16. For an example of a pure research launch see: Memorandums (2), Ernest J. Ott for the Files, 20 September 1962, both in folder 005085 "15.6 Journeyman," in file tray "Sounding Rockets General (con't) Alphabetical Aerobee thru Javelin," in NHO. This was a Univ.of Michigan radio astronomy launch. For an applied research project of Ohio State University, with Air Force sponsorship, see: Letter, Ross Caldecott to E. A. Brummer, 7 August 1963; Letter, T. R. Patterson to Mr. Littell, 5 September 1963, both in folder "Special File, May 63 - February 64," in RGA181-l(S).

17. "Robbins," OHI, Tape lb: 445. Joseph Robbins notes that problems arose when scientists would move to different universities and attempt to take their grant money along. The grant, having been made to the given institution, stayed with that institution wherever the research went.

18. Ibid.

19. Ibid. "Milliner," OHI, Tape la: 505, for quotation; lb: 420 for education.

20. Letter, Robert L. Krieger to Director, Langley Research Center, 20 May 1963; Letter, U. M. Staebler to Floyd Thompson, 7 June 1963, both in folder "Special File, May 63 - February 64," in RGA181-l(S). See also: *A&A, 1963*, 208.

21. Letter, Robert L. Krieger to Langley Research Center, 20 September 1963, in folder "Special File, May 63 - February 64," in RGA181-l(S).

22. Letter, J. W. Wright to S. L. Seaton, 10 June 1964 in folder "Special File, March - December 64," in RGA181-l(S).

23. "Spinak, et al.," OHI, Tape 2a: 440-80; "Milliner," OHI, Tape lb: 250.

24. Letter, F. W. Reichelderfer to Director Langley Aeronautical Laboratory, 15 September 1958, in folder "Special Files, September December 58," in RGA181-l(S); the 17 October reply is also in this file. See Shortal' 544, for a radar image of the hurricane.

25. *Data Book II*, 346-48. Letter, Earle F. Cook to Director Langley Research Center, 12 January 1959, in folder "January - May, 1959," in Wallops box #4.

26. Memorandum, H. R. Brockett for Dr. Tepper, 18 October 1960, in folder 005065 "15.1 Sounding Rockets to 1963," in file tray "Sounding Rockets - General," in NHO. Letter, Robert L. Krieger to Willis L. Webb, 21 December 1961, in folder "Wallops, January - March 62," in RGA181-l(C). The "NASA ARP, April 1961," page arp 8.2, lists other firing sites as Pt. Mugu (CA), Eglin Field (FL), White Sands, and Tonapah (NV).

27. Letter, Robert L. Krieger to Morton J. Stoller, 9 April 1962, in folder "Special File, May - September 62," in RGA181-l(S).

28. Letter, Robert L. Krieger to Morton J. Stoller, 24 May 1962, Ibid.

29. Airmail, Herbert A. Wilson Jr. to E. Whitney, 14 March 1963, in folder "Special File, October 62 - April 63," in RGA181-l(S); Memorandum, Harold N. Murrow for Associate Director Langley Research Center, 30 April 1965, in folder "Wallops, January - June 46 [sic]," in RGA181-l(C).

30. Letter with enclosure, Robert L. Krieger to Director of Meteorological Programs, 15 December 1965, in folder "Special File June - December 65," in RGA181-l(S).

31. "NASA APR, March 1961," page sfp 27.5.

32. Shortal, 543, for the FPS-16 radar; "NASA APR, June 1962," page A 40.0 for NASA contract NAS6-386 to RCA, "Services necessary to operate and maintain the Wallops Station AN/FPS-16 radar and Tiros ground data acquisition systems." Contract was for $216,000.

33. "NASA APR, April 1961, page osfd 1.2, for the transfer of the station. Memorandum, Edmund C. Buckley for Director Space Flight Programs, 10 July 1961, in "Chron. File, July - December 1961," in NASA HQ box #1. "NASA APR, December 1961" page B 1.4, for station overhaul.

34. Pamela Mack, "Satellite and Politics: Weather, Communications, and Earth Resources," in *Spacefaring People*, Roland, ed., 32-34, for NASA-Weather Bureau controversy.

35. "NASA APR, September 1962," page B 1.1, for Tiros, B 2.2, for Nimbus.

36. Letter, Gerald M. Truszynski to Homer E. Newell, 21 September 1964, in folder "July thru December, 1964," in NASA HQ box #1.

37. See chapter 2, above, for Project 2080. U.S., Congress, Senate, Committee on Astronautics and Space Sciences, *NASA Authorization for 1960, Hearinqs before a Senate subcommittee of the Committee on Astronautics and Space Sciences on S. 1582 and H.R. 7007*, 86th Cong. 1st sess., 5905-21S, 792, for FY 1960 Wallops funding wherein Administrator Glennan told the Senators that the build-up at the island was "essentially complete." U.S., Congress, House, Committee on Appropriation, *Independent Offices Appropriations for 1961, Hearings before a House subcommittee of the Committee on Appropriations*, 86th Cong. 2nd. sess., 6003-OlH, 353-55 for FY 1961 funding. See also: *Data Book I*, 491, table 6-147 for "Funding by Fiscal Year;" and 168, table 4-28 for "Construction of Facilities Direct Obligations, by Installation." The figures in these charts differ somewhat due to the fact that the money requested in a given year might be spent over several.

38. U.S., Congress, House, Committee on Science and Astronautics, *NASA Authorization for 1963, Hearings before the House Committee on Science and Astronautics on H.R. 10100*, 87th Cong. 2nd. sess., 6203-06H, 944, 948. U.S., Congress, House, Committee on Appropriations, *Independent Offices Appropriations for 1963, Hearings before a House subcommittee of the Committee on Appropriations*, 87th Cong. 2nd sess., 6203-13H, part 3, 882-87. By comparing the figures from these two hearings I noticed that Wallops figures on page 948 of the former are misprinted, someone copied the number above the summation line instead of the total.

39. *Data Book I*, 489, table 6-145 "Personnel".

40. "Spinak, et al.," OHI, Tape 2a: 545. This 50/50 mix of contract/in house personnel was unusual compared to NASA's average of 3 to 4 contract personnel to each NASA employee during this time. See: *Data Book I*, 118, table 3-26 "Total NASA Employment, Selected Characteristics."

41. *Data Book I*, 345, table 6-54 "Funding by Fiscal Year," for the Cape's budget; 415, table 6-94 "Personnel," for employment at Houston.

42. "Robbins," OHI, Tape Ib: 300-20.

43. Memorandum, Abe Silverstein for Directors, 6 April 1960, in folder "Wallops, January to June 1960," in RGA181-l(C).

44. Cover Letter, Robert L. Krieger, 8 December 1961, in folder "Wallops Station Handbook, DTD 12-8-61," in RGA181 l(C). This folder contains a copy of all four volumes of the Handbook. I: General Information; II: Flight Test and Support Facilities; III: Range Users Information; IV: Safety. Letter, Robert L. Krieger to Langley, Goddard, and Lewis Centers, 8 January 1962, in folder "Special Files, January - April 62," in RGA181-l(S), for 1 January effective date for the Handbook's procedures.

45. "NASA APR, October 1960," page lvp 51.3; "January 1961," page lvp 51.3; "April 1961," page lvp 51.3, follow the course of the decision on this issue.

46. Memorandum, Paul A. Price for File, 25 September 1961, in "Chron. File, July - December 61," in NASA HQ box #1. This memo describes the "September Meeting of the DOD Inter-range Communications Planning Committee." Price lists 3 representatives from AMR, 2 from PMR, and 1 each from White Sands, Defense Communications Agency, and NASA (Price himself). No one from Wallops seems to have attended. Recall, also, that the investigations of labor problems at missile bases ignored Wallops. In "Spinak, et al.," OHI, Tape 2a: 380, Engineer Robert Duffy states that Wallops "has always been an Associate Member," of the DOD's "Range Commander's Council," but it seems that most of the contact between Wallops and the other ranges took place on an informal basis in the early days.

47. Letter, Edmund C. Buckley to Brigadier General Paul T. Cooper, 4 May 1962, in "Chron. File, January thru December 1962," in NASA HQ box #1. Memorandum, Andrew G, Swanson for Associate Director, 10 July 1962, in folder "Special File, May - September 62;" Memorandum, Carl A. Sandahl for Associate Director, 10 December 1962, in "Special File, October 62 - April 63," both in RGA181-l(S).

48. "NASA APR, May 1962," page D 2.4, for Scout pad refurbishment. For increase in Scout capacity see: Memorandum, Floyd L. Thompson to Wallops, 27 September 1962, in folder "Special Files, May - September 62," in RGA181 l(S).

49. Levine, 183. Levine notes that Congress tightened-up NASA's ability to use this technique in 1965. See also: "NASA APR, September 1962," page A 4.4; U.S., Congress, Senate, Committee on Aeronautical and Space Sciences, *NASA Authorization for FY 1964, Hearings before the Senate Committee on Aeronautical and Space Sciences on S. 1245*, 88th Cong. 1st sess., 6304-24S, 67.

50. U.S., Congress, Senate, Committee on Aeronautical and Space Sciences, *NASA Authorization for FY 1964*, S. Report 385, 88th Cong. 1st sess., 6308-02S, 331-2, for quotation. For NASA planning for this budget see: Report, "Wallops Island Fiscal 1964 Estimates Supporting Facilities," 19 December 1962, in "Chron. File, January - December 1962," in NASA HQ box #1. U.S., Congress, House, Committee on Science and Astronautics, *NASA Authorization for 1964, Hearings before the House Committee on Science and Astronautics on H.R. 5466*, 88th Cong. 1st sess., 6303-12H, part 4, 2897-2898. While I have no direct evidence that this denial of funding was in retaliation for the reprogrammed bypass of normal channels, it should be noted that the Senate Authorization Committee was usually the most lenient committee NASA had to appear before, and Wallops tried again for the funding (apparently without success) in fiscal 1966.

51. U.S., Congress, House, Committee on Aeronautical and Space Sciences, *Authorizing Appropriations to the National Aeronautics and Space Administration*, H. report 1674 to accompany H.R. 11737, 87th Cong. 2nd sess., 6205-15H, 145; U.S., Congress, Senate, Committee on Aeronautical and Space Sciences, *NASA Authorization for FY 1963, Hearings before the Senate Committee on Aeronautical and Space Sciences on H.R. 11737*, 87th Cong. 2nd sess., 6206-13S, 745.

52. The Aerobee is a small, liquid-fueled sounding rocket, one of the few liquid-fuel vehicles Wallops handles. For operational status of the tower see: "NASA APR, February 1960," page sfp 65.3.

53. Memorandum, Charles J. Donlan to NASA - Code RTM, 30 June 1961, for first quotation. Memorandum, Abe Silverstein to Director Advanced Research Programs, 9 August 1961, for Silverstein's quote, emphasis in original. Both memos are in "Chron. File, July - December 1961," in NASA HQ box #1.

54. Memorandum, T. Melvin Butler to Robert L. Krieger, 17 May 1965, in folder "Wallops, January - June 46 [sic]," in RGA181-l(C).

55. U.S., Congress, House, Committee on Science and Astronautics, *NASA Authorization for 1964, Hearings before a House subcommittee of the Committee on Science and Astronautics on H.R. 5466*, 88th Cong. 1st sess., 6303-12H, part 4, 2890, for Buckley's testimony. See also: Memorandum with enclosure, W. J. Boyer to R. L. Rrieger, 9 May 1962, in folder "Special File, May - September 60," in RGA181-l(S); "Spinak, et al.," OHI, Tape 2b: 330-80.

56. "Spinak, et al.," OHI, Tape 2b: 330-80. *Data Book I*, 485, table 6 141. There is a photo of *Range Recoverer* in *A&A*, 1966, 186.

57. Letter, Robert L. Krieger to NASA Headquarters (Code T), 25 September 1964, in folder "Special File, March - December 64," in RGA181-l(S). *Data Book I*, 485, table 6-141. C-121 was the military designation for the Lockheed Constellation passenger airliner. See also: "Floyd," OHI, Tape la: 250.

58. Memorandum, G. T. Ragon to Commander Service Squadron 8, 6 August 1962, in folder "Special File, May - September 62," in RGA181-l(S), for Navy diver service. Memorandum with enclosure, Abe Silverstein for Associate Administrator, 14 April 1961, in "Chron. File, January - June 61," in NASA HQ box #1, for "NASA request to CNO to extend Navy support of Wallops Island Operations." U.S., Congress, House, Committee on Science and Astronautics, *NASA Authorization for 1963, Hearings before the House Committee on Science and Astronautics on H.R. 10100*, 87th Cong. 2nd sess., 6202-27H, part 1, 118, for funding for naval support.

59. Letter, Floyd L. Thompson to Commanding Officer ZW-l, 11 January 1961; Letter with enclosures, Floyd L. Thompson to Wallops Station, 11 January 1961, both in folder "Special Files, January - March 61," in RGA181-l(S).

60. *National Aeronautics and Space Act of 1958*, Section 102, c(7).

61. Don E. Kash, *The Politics of Space Cooperation* (West Lafayette, IN.: Purdue Research Foundation, 1967), 1-23.

62. Ibid., 50-76. The four areas of cooperation are: information exchange, personnel exchange, operations support, and cooperative projects. See also: Draft #2, "A statement of responsibilities of the Director of the Office of International Programs," 13 February 1959, in folder 11.2 "International Affairs, Office of," in box "Office of International Affairs; Office of U.N. Conference (OUNC); Office of Policy Coordination and International Affairs," in NHO. Hereafter cited as "folder 11.2 in box OIA." Logsdon, "Opportunities.

63. "Milliner," OHI, Tape la: 520-lb: 50.

64. Roland, I: 4-5. For Canadian and NATO projects at Wallops see chapter 1 above.

65. Green and Lomask, 18-26.

66. "Milliner," OHI, Tape la: 190; "Spinak, et al.," OHI, Tape 2a: 195, for Buckley's relationship with Wallops. *Data Book II*, 546, for MSFN negotiations.

67. Statement, "International Programs of the National Aeronautics and Space Aaministration," 9 February 1961; Memorandum from the Administrator, T. Keith Glennan, 1 April 1959, both in folder 11.2 box OAI. See also: *Data Book II*, 524.

68. "NASA APR, December 1959," page OIP 17.2. For the CF-105 program see chapter 1 above.

69. U.S., Congress, House, Committee on Aeronautics and Space Exploration, *International Cooperation in the Exploration of Space*, H. Report 2709, 86th Cong. 1st sess., 5901-03B. Statement, Hugh L. Dryden before the Subcommittee on Independent Offices of the Senate Committee on Appropriations, 19 May 1960, 13, in folder "VIII Budget: Background Briefings, Supplemental Material (FY 1961)," in file tray "Budget, FY 1961, FY 1962," in NHO. Aside from noting that "sounding rockets and Scout satellites," were functions of Wallops, the reference to international cooperative projects provides the bulk of the Wallops mission definition that Dryden gave the Committee.

70. "NASA APR, July 1960," page IPP 52.3, for meeting. *Data Book II*, 291-93 for Ariels 1, 2, and 3; *Data Book III*, 184 85, for Ariels 4 and 5.

71. The NASA APR gives a good running record of foreign researchers at Wallops each month. For examples see January 1961, page GA 17.4 for French professor, "utilizing specialized optical equipment," and page 17.6 for the Norwegian engineer, "studying range telemetry systems.'

72. "Milliner," OHI, Tape la: 520.

73. "Rockets' Red Glare Lights Up Remote Island," *The Washington Post*, 9 August 1980, B7, first quote is from Joyce Milliner, second is from Robert Krieger.

74. Public brochure, Robert L. Krieger, undated (for Open House 28-29 September 1963), in folder 004680 "Wallops General (1958-1963)," in file tray "Centers, Wallops Flight Facility," in NHO.

75. "Robbins," OHI, Tape la: 535.

76. "Spinak, et al.," OHI, Tape 2b: 250. *Data Book I*, 482, indicates that equipment was on loan to India, Pakistan, Argentina, Brazil, and Spain. See also: Library of Congress Staff Report, "Meteorological Satellites," for the Senate Committee on Aeronautical and Space Sciences, 6203-29S, 112, for the November 1961 visit of delegates to the "International Meteorological Satellite Workshop," to Goddard and Wallops.

77. Minutes, Administrator's Staff Meeting, 18 January 1961, 3, in book #2, in box "Administrator's Staff Meeting Minutes, October 60 - June 61," for an account of the Sardinia launches, "Two Wallops people were there." For Italian interest in Scout see: "NASA APR, October 1961," page A17; "NASA APR, April, 1962," page A 17.1.

78. U.S., Congress, House, Committee on Science and Astronautics, *Proceedings of the Panel on Science and Technology, 6th Meeting, before the House Committee on Science and Technology*, H. Report, 89th Cong. 2nd sess., 6501-26H, 13, for the statement, "European Progress in Aeronautics by Professor Luigi Broglio, President Italian Space Research Commission."

79. Centro Ricerche Aerospaziali of Roma, Italy, *San Marco Range*, undated (early 1972) leaflet, in folder "FLT/Scout," in FLT Papers. See also: *Data Book II*, 64-65, 299-300.

80. Letter, Robert F. Garbarini to Langley Research Center, 26 August 1965; Letter, Sidney Teweles to Morris Tepper, 17 August 1965; Letter, Robert F. Garbarini to Wallops Station, 26 August 1965, all in folder "Special File, June - December 1965," in RGA181-l(S).

81. Minutes, and "NASA APRs" as cited in note 77 above. A sodium vapor rocket dispensed the chemical at a programmed altitude creating a very visible cloud. The motions of this cloud provided data relating to atmospheric motions.

82. "Robbins," OHI, Tape lb: 430. Memorandum, Edmund C. Buckley for Dr. Silverstein, 6 October 1961, in "Chron File, July - December 1961," in NASA HQ box #1. *Data Book I*, 487, table 6-143.

83. "Milliner," OHI, Tape la: 530.

84. "NASA APR, April 1961," page LVP 51.3, for 14 April meeting. "NASA APR, August 1961," page LVP 51.4, for the reversal of the Scout transfer. Vought Corp. was the prime contractor for Scout.

85. Memorandum, Floyd L. Thompson to Robert L. Krieger, 6 March 1962, in folder "Special Files, January - April 62," in RGA181-l(S). NASA Wallops News Release #63-12, "Wallops Station Personnel Participate in West Coast Launching," 12 February 1963, in folder 005085 "15.6 Journeyman," in file tray "Sounding Rockets General," in NHO. Letter, Floyd L. Thompson to Robert L. Krieger, 21 May 1963, in folder "Special File, May 63 - February 64," in RGA181-l(S).

86. Memorandum, R. B. Morrison to SD/Deputy Associate Administrator, 14 April 1964, in folder 006255 "Documentation OSO," in file tray "Earth Satellite Probes, Open - OSO," in NHO. Minutes, Administrator's Staff Luncheon, 22 April 1964, 5-6, in folder #4 "January - November 1964," in box "NASA Administrator's Policy Meetings, (1962-1965), Action Items (1962-1968)," in NHO. *A&A, 1964,* 135, 141, 147, 166. The fatalities were Sidney J. Dagle and L. D. Gable of Ball Brothers Co., and John W. Fassett of Goddard.

87. "Spinak, et al.," OHI, Tape 2a: 382. Letter, Robert L. Krieger to List, 19 June 1964, in folder "Special File, March - December 64," in RGA181-l(S).

88. "Report of accident at Wallops Island on June 8, 1964," Hubert K. Clark, 12 June 1964, in folder "Special File, March - December 64," in RGA181-l(S). "Contributing factors," noted are: "a. Too many organizations making decisions without proper designation of single point of authority; b. Requirement for meeting tight schedules on a non-interference basis; c. Inadequate review of technical requirements of proposed systems."

89. "Spinak, et al.," OHI, Tape 2a: 400-20, first two quotes are from Abraham Spinak, last quote ("and assume that...") is from Robert Duffy.

90. Shortal, 193-96, for accident involving Durwood A. Dereng. Spinak refers to this accident in OHI, 2a: 400. For Krieger's vehemence on range safety see letter cited in note 87, above.

91. "NASA APR, July, 1962," page D 2.4. It is interesting to note that, "Wallops has requested Langley review the ARGO D-8 aerodynamics and structure. GSFC has supplied Langley with a number of reports on the ARGO D-8 vehicle."

92. Memorandum with enclosure, G. M. Truszynski for Associate Administrator for Space Science and Applications, 24 June 1964, enclosure page 2, in folder 005064 "15.1 Sounding Rockets 1964-69," in file tray "Sounding Rockets General," in N80. DEW is Defense Early Warning, a series of radar sites to detect incoming missiles; NBS is the National Bureau of Standards; W.B. is the Weather Bureau.

93. Memorandum, Facilities Engineer to E. C. Buckley, 16 October 1964, Ibid. "Milliner," OHI, Tape lb: 525.

94. Memorandum, Ibid, for ARL support. "Milliner," OHI, Tape lb: 525, for travel budget. "Robbins," OHI, Tape lb: 420, for logistics. See also: *Data Book I* 482 table 6-138.

95. Staff Study, "An Accelerated NACA Space Flight Program (Langley Version)," undated (early 1958), in folder "NASA - Space Flight Program 1958," in FLT Papers. This study includes "cost breakdowns" for: "Conversion of each of 3 ships for down-range tracking and instrumentation and for the launching of small (100 lb.) satellites," and "modifying and equipping one large ship for launching and monitoring flight of manned satellite."

96. "The Sun's Awesome Impact," *Life,* 28 November 1960, 78, for a photo of Navy ship launching eclipse research rocket.

97. U.S., Congress, House, Committee on Science and Astronautics, *Equatorial Launch Sites - Mobile Sea Launch Capability*, H. Report 710, 87th Cong. 1st sess., 6107-12H, quotation is on page 6.

98. Memorandum, H. R. Brockett to E. P. Odenwalder, 15 August 1961, in "Chron. File, July - December 1961," in NASA HQ box #1. See also in the same file: Memorandum, E. C. Buckley to H. J. Goett, 5 December 1961, on the same subject. Rawinsonde [RAdar-WINd-radioSONDE] operations entailed using the ship's tracking and telemetry equipment to support weather balloon launches, some of which launched from the ship, some from shore.

99. "Spinak, et al.," OHI, Tape lb: 170.

100. Memorandum, Robert P. Rhinehart for Associate Director, 11 December 1962, in folder "Special File, October 62 - April 63," in RGA181-l(S). Memorandum with enclosure, Hugh S. Fosque for Range Engineering Branch Files, 5 February 1963, in folder 005105 "Mobile Launch of Sounding Rockets from Shipboard (1963-64), in file tray "Sounding rockets (con't) Prometheus thru Wasp."

101. Ibid.

102. "Spinak, et al.," OHI, Tape lb: 180, first part of quote is from Robert Duffy, second part (and they bought it..) is from Abraham Spinak.

103. National Academy of Sciences, *International Geophysics Bulletin #69*, March 1963, "Provisional Programme International Years of the Quiet Sun, 1964-65," in folder OI 391400-01 "International Q. S. Year," in Space History Collection, NASM. Wallops participated in the IQSY with several launches, including *Explorer XXX* on 19 November 1965. In regard to the solar cycle, it should be noted that the polarity of sunspot pairs reverses with each 11 year cycle, so that astronomers usually speak of a 22 year cycle. A given source might refer to either.

104. Letter, John E. Naugle to Stanley Ruttenberg, 21 November 1963, in folder 005105 "Mobile Launch of Sounding Rockets from Shipboard (1963-64), in file tray "Sounding rockets (con't) Prometheus thru Wasp," in NHO.

105. Memorandum, Harold N. Murrow to Associate Director, 7 April 1964, in folder "Wallops, April - June 64," in RGA181 l(C). This memo states, "The Sandia Corporation has a requirement for high altitude wind measurements near the equator both for fallout studies and other classified projects." See also: "Carrier to Serve as Rocket Launch Range," *Aviation Week and Space Technology*, 19 October 1964, 33.

106. Memorandum, Robert L. Krieger to Floyd L. Thompson, 29 June 1964, in folder "Special File, March - December 64," in RGA181-l(S).

107. Memorandum, Robert L. Krieger to Floyd L. Thompson, 28 July 1964, in folder "Wallops, July - December 64," in RGA181-l(C).

108. Escort carriers were small ships with a limited number of aircraft that provided air cover for convoys, protecting them primarily from submarines. *Croatan's* small size combined with its designed ability to handle aircraft made it a suitable platform for sounding rocket operations.

109. Memorandum, Abraham Spinak to Floyd L. Thompson, 19 November 1964, in folder "Special File, March - December 64," in RGA181-l(S). See also: *A&A, 1964*, 384-5, 391, 393.

110. NASA Wallops News Release #64-86, "Sounding Rocket Ship Completes Shakedown Cruise," 23 November 1964, in folder 005104 "Sounding Rockets IQSY Program, 1964-65 (also USNS Croatan)," in file tray "Sounding rockets (con't) Prometheus thru Wasp," in NHO.

111. "Spinak, et al.," OHI, Tape lb: 157-240. In folder 005106 "Shipboard Sounding Rockets," in NHO, there are a number of photos of the preparations for this voyage. In folder 005104 "Sounding - Rockets IQSY Program," photo G-65 4787 shows *Croatan* in Baltimore Harbor. Interestingly this photo was issued by the Goddard public information office.

112. *A&A, 1965*, 110, 121.

113. "Robbins," OHI, Tape lb: 390. NASA Wallops News Release #65-22, "NASA Sea-Going Platform Completes Launch Expedition," 22 April 1965, in folder 005104 as cited in note 110 above.

114. *A&A, 1965*, 163. NASA Wallops News Release #65-23, "NASA Sea-Going Expedition to Sail in Mid-February," 11 February 1965, in folder 005104, as cited in note 110 above.

115. i, 169. Chart, "Launch Schedule for Mobile Launch Facility Expedition No. 1," undated, in folder 005106 "Shipboard Sounding Rockets," in NHO. This pre-expedition schedule is not completely accurate for retrospective dates, but is useful in showing the general itinerary of the mission. See also: NASA Wallops News Release #65-23, as cited in note 114 above.

116. NASA Wallops News Release #65-22, as cited in note 113 above. Report, Robert L. Krieger to Langley Research Center, 26 November 1965, in folder "Wallops, July - December 65," in RGA181-l(C). This report lists the following numbers of major rocket launches during the mission (keeping in mind that small rockets were launched on virtually a daily basis): Langley (18), Univ. Michigan (14), Goddard (13), Naval Ordinance Test Station (9), Univ. New Hampshire (8), Sandia Corp. (7), Air Force Cambridge Research Lab (6), Univ. Illinois (5).

117. "Spinak, et al.," OHI, Tape 1b: 200.

118. Ibid., Tape 2b: 360. "Milliner," OHI, Tape 1b: 545. *A&A, 1966*, 87, 178, 185-87.

Chapter 5

CHANGES AMID CONSTANCY

The flight facility that "sold" the Mobile Sea Launch Expedition to NASA in 1964 differed noticeably from the test station of the NACA era. The most visible difference was the growth of the physical plant. The Pilotless Aircraft Research Station consisted of a couple of simple launch pads, a few buildings, and World War II surplus radar equipment located on one half of a small island, and a minor number of diminutive tracking sites downrange. In the span of seven years Wallops Station's resources grew to include five launch areas, an airfield, sophisticated rocket handling and check-out facilities, and expensive tracking and data acquisition equipment, while encompassing three separate locations at the Virginia site.[1] Wallops personnel no longer conducted operations solely from the island and for select customers; they ran or oversaw facilities in Alaska, Bermuda, Manitoba, and North Carolina, as well as aboard various ships, and provided services to a wide range of customers foreign and domestic.

The acceleration of the American space program brought a steady increase in the number of NASA employees assigned to the base, and the introduction of a significant number of contract personnel for both research and support, essentially quadrupled the overall size of the workforce within the same period. The shift of the Station's administrative functions from Langley, and the incorporation of the NACA in NASA, resulted in new missions and more challenges for the staff. The higher level of public interest and visibility forced the staff to establish policies and procedures to deal with both the observers and the new range users attracted to the base. Despite the occasional miscue, the public, Congress, and Headquarters generally supported the Station, and its position within NASA and the space program solidified.

The main source of the changes at the base stemmed from the difference in the nature of the old aeronautical research as opposed to the new space endeavors. NASA was not just the NACA renamed. The elder institution focused on a narrow facet of engineering research, with some (largely unwanted and politically inspired) developmental work mixed in. NASA, to be sure, conducted such engineering research, but the new organization gave as much weight to developmental work and science research and spread its resources out over a broader range of ventures.[2] Also, the inclusion in NASA of groups and institutions that had cultures and operational philosophies different from the NACA served to fundamentally alter the style of the space agency. The perceived importance of the space program to the national interest, by most of the public and political leadership, led to an infusion of funds and a sense of purpose rare in peacetime.

All this having been said, an old adage comes to mind: the more things change, the more they stay the same. Even with all of the transformations to the mission and make-up of the base, many features of the original facility survived the transition era. As the focus of the research conducted at the base moved from transonic through hypersonic to space research, the variety and complexity of the equipment, and the sheer number of experiments, increased apace. In the face of this variety and complexity Wallops' raison d'etre remained the launch, tracking, and acquisition of data from small rockets and balloons using radar and radio-telemetry techniques. These projects could have been, and often were, conducted from other facilities, but Wallops remained a part of most of them so that the larger, more complex launch facilities at the Cape and Vandenberg could concentrate on the high-priority, expensive missions, and not suffer interference from the low-budget, unspectacular portion of NASA's program.

The value of Wallops as a civilian owned range increased as NASA's mission became more publicized, and open to international participation. And even while it carved out an independent administrative niche for itself amid NASA's changing goals and internal organization, the Station continued to be insulated, able to operate with an "informal procedure that was formalized."[3] Personal contacts and informal discussions continued to be the common procedure for budget and research planning throughout the period.

Regardless of the expansion of the base and its increased visibility, Wallops remained a relatively isolated NASA outpost. The surrounding area grew only slowly during this period, and certainly differed from Cleveland, Houston, Washington, or the other metropolitan areas that hosted the majority of NASA facilities, in terms of accessibility, population, and economic resources. The distinctly rural flavor of the area was not for everyone. "I know sometimes we had trouble recruiting engineers and scientists because you either have to like a rural area or, you know, you're just not happy. And we did have some that came, and their families didn't like it and they left; but most of them came and liked it and stayed."[4] Though the Station was integrated into the NASA communications network, could be easily reached by air, and played host to visiting researchers from all over the world, the nature of its environs played a major role in shaping the composition of the permanent staff.

During this formative era NASA reorganized several times. The first reorganization occurred in 1959 in response to the absorption of the Army Ballistic Missile Agency group. As discussed earlier in chapter two, this reorganization caused a brief debate between the Office of Space Flight Programs and the Office of Launch Vehicle Programs over the position of Wallops within NASA. In 1961 the structure of the space agency was altered in response to both the installation of James E. Webb as Administrator (he replaced T. Keith Glennan in February), and the announcement of the lunar landing goal in May. The effects of this reorganization included the abolition and restructuring of the old program offices (including both OSFP and OLVP), the creation of a separate Office of Tracking and Data Acquisition (under Edmund Buckley), and the situation of the field centers directly under the control of Associate Administrator Robert C. Seamans. Wallops and the other field centers thus had coequal status within NASA and, theoretically, could bring their problems, proposals, and requirements directly to the attention of upper management.[5] The idea was to remove the layers of insulation that existed between the top levels of Headquarters and center managements, make it easier for the centers to perform multifaceted tasks, and ease coordination between related projects such as spacecraft and launch vehicle development. The centers still received "program direction" from the new Headquarters program offices, but received funding and support through Seamans.[6] Also, Webb wanted a "flexible organizational and administrative framework," for the acceleration of the space program.[7]

Unfortunately, this plan did not work out as expected. The centers were too many and their work too diverse to allow Seamans to adequately oversee all of them. Center directors could not get sufficient access to the Associate Administrator, and they and the program directors could not efficiently coordinate their efforts.[8] It also seems that the scale and speed of the Apollo program exceeded expectations. In June 1962 "a meeting involving the field administrative officials and selected headquarters representatives," was organized and scheduled to take place at Wallops. Director of Administration Albert F. Siepert selected the site because it was one "which most of you have not visited."[9] This meeting and other "adjustments" could not sufficiently remedy the situation and in November 1963 NASA's structure again shifted. This time the field centers returned to the control of various Headquarters program offices, and Wallops again found itself grouped with Goddard and JPL under the space science section of NASA, Homer Newell's Office of Space Science and Applications.[10]

These administrative shuffles had little effect on the day-to-day operations at Wallops. Being a "service center," the base operated differently from most of the other field centers. Wallops engineers did not manufacture payloads; they launched, tracked, and sometimes recovered equipment brought to them. They trained launch and tracking crews and participated in many projects, but until the *Croatan* operations in 1964-65, they played a support rather than a leading role in these projects. Therefore, these reorganizations, "didn't make a who[le lot of difference.] I knew the people, ..., you had to sort of learn how they operate. You know, when you make a new organization like [OSSA], if you can get in there first and let them know you want to work with them, and all, they feel happy about this so it sort of greases the way for later on."[11] Even under the 1961 structure, budget and operational planning at Wallops involved contact with different parts of the Headquarters organization.[12]

The multiple reorganizations of NASA's Headquarters structure stemmed in part from the youth of the agency. The internal structure at Wallops, an older, more well established facility, remained relatively stable during this period. The two changes of note which did occur involved adjustments to meet the changing priorities at the base, rather than the need for any wholesale administrative renovation. The Range Engineering Branch, originally a part of John Palmer's Flight Test Division, became a separate Division in response to the increased variety of the tracking and data acquisition programs underway at the base, and as a reflection of the separate status of Buckley's OTDA in Headquarters. Secondly, a Program Management and Liaison Branch, placed within the Flight Test Division and headed by Cary F. Milliner, served to coordinate the many projects coming to the range from its diverse customers. An examination of Wallops' organizational charts reveals, however, a continuity in the personnel occupying the top positions at the base. The Division Chiefs and most of the Branch Heads listed in a March 1961 chart are still listed in June 1967. This continuity provided stability at the base throughout a hectic period and contributed to the development and propagation of a Wallops culture at the Station.[13]

The continuity and stability at Wallops proved fortunate as Krieger and his staff soon had to deal with an unforeseen administrative headache. The rapid pace of the space program's expansion brought forth the issue of how efficiently the expansion of the various NASA facilities was proceeding. "The objective of orderly introduction of uniform NASA standards in the construction of facilities was expressed by the House Committee on Science and Astronautics in the NASA Authorization Act for the fiscal year 1963."[14] In October 1964 the Committee notified Administrator Webb that, "For its information and guidance, the committee is undertaking a review of Federal Government policy regarding the planning of base facilities required for the space program. The NASA field installations, of course, are of particular interest."[15]

The Committee desired to learn how well "master planning" at the field centers adhered to accepted government procedures. This included examination of the status and maintenance of planning documentation, timing and requirements for such planning at various bases, and the relative efficiency of the agency-wide expansion at those bases.[16] The Committee first examined the planning practices of the Army, Air Force, General Services Administration, and the Atomic Energy Commission, as "representative of the standards and criteria applied in the Federal Government."[17] Then they began to visit NASA field centers.

Goddard Space Flight Center, "influenced" by its relationship with the National Capital Planning Commission, possessed master planning documentation and activities the Committee found acceptable. Marshall

Space Flight Center had the planning inherited from its Army origins. The Kennedy Space Center shared its planning tasks with the neighboring Air Force base and several companies under contract. Five other facilities also passed muster with varying grades. Two, however, did not. Lewis Research Center and Wallops Station drew fire from the Committee, which noted that these two, "avoid [master planning] with disdain."[18]

The Committee recognized that Wallops consisted of an amalgamation of NACA, Navy, and NASA construction, but were unsympathetic. "The absence of a master plan is deliberate. Two reasons are given: (1) Wallops, it is explained, doesn't really have a program of its own; it implements the programs of other NASA Centers and, therefore, must wait upon their decisions before being able to identify facility needs. ... (2) Rapid changes in technology require flexibility of development which, in the view of a staff officer, would be circumscribed by a long-range master plan. The station policy on facility planning and construction is summed up in this official's concluding observation: 'We build no monuments here.'"[19]

The Committee report argued that given the fact, "many buildings on the island and at the main base appear to be approaching structural obsolescence, ..., the view advanced at Wallops that facility planning must await authorized development is open to question." Continuing expansion at the base, and a need to "convert, demolish, or replace," many facilities presented a need for master planning."[20] It concluded that the process of master planning was economical, "as illustrated by the confused and congested layouts at Lewis and Wallops."[21] While conceding that NASA Headquarters had not, until prodded by the Committee, paid much attention to this issue, they stated that master planning needed to be an integral part of the operation, "at *all* NASA installations."[22] Needless to say, Krieger got the hint.

On 3 August 1965, Wallops contracted for $33,400 to American Engineers, of Richmond, for "services and materials for a master site plan."[23] It was beneficial for the Wallops staff to move quickly because the representatives of the House Committee revisited the base later that month in the course of preparing a follow-up report.[24] This time the Committee found, "There has been a substantial change since December 1964 in the Wallops Station policy and practice of master planning." They reported that the staff had reviewed several older studies done by Langley, the Navy, and private companies, and followed with the preparation of a three phase plan.[25] The report noted that, "Obviously the Wallops Station management in the last 8 months has taken a long step forward in the direction of master planning of the base facilities. ... The new Wallops planning program, in its three phases, appears to be well conceived."[26] It also praised Headquarters, Lewis, and Wallops, "for modifying contrary policies of a year ago."[27]

Krieger and company had no desire to antagonize Congress. The death of Representative Albert Thomas in February 1966 and the general success of the Wallops program did not lessen Congressional scrutiny. Thomas'

111

successor as chair of the Independent Offices Appropriations Subcommittee, Joseph Evins, questioned the need for Wallops and the sounding rocket program in 1966 hearings but did not press the issue in the face of strong support given by Webb and Newell.[28] In spite of questions of this nature, Wallops did not receive a serious challenge to its existence during this time. The margin of safety provided by being clearly ensconced in the space program was demonstrated by the House attempt, in 1963, to close the NASA Flight Research Center at Edwards Air Force Base. Similar in size, budget, and complement to Wallops, the California facility had for years hosted research aircraft from the X-1 to the X-15. The House, operating under the perception that a facility dealing most visibly with aeronautical research constituted an anachronism in the space age, proposed closing the base to save money. Quick work by NASA saved the base, but Wallops apparently took little notice of the plight of the future Shuttle landing facility.[29]

The important factor was the general decline in the status of aeronautical research in NASA. While it continued to be a part of the agency's program, there is no question that it received a decidedly lower priority than the space programs, especially when Project Apollo appeared.[30] Indeed, some aerodynamicists, like Langley veteran John Stack, left NASA to pursue research elsewhere.[31] While Wallops' program continued to include some aeronautical projects, principally through the use of the runways, its primary task had become support of space science. This identification with "new" space research, as opposed to "old" aeronautical research, staved off threats to the installation's existence until the novelty (and perceived importance) of space research waned in the 1970's.[32]

The three year span of 1964-66 turned out to be a high-water mark for NASA. Though the lunar landings still lay in the future, funding and employment levels began to drop. The completion of major construction of Apollo infrastructure and the award of contracts for flight hardware marked the end of the massive space expenditures of NASA's early era. The Johnson Administration's guns and butter approach to the dual commitments of Vietnam and the Great Society began to absorb more of the federal budget. Also, Apollo became the focus for the whole of NASA, and other programs received less attention.[33] As it fulfilled President Kennedy's challenge, the agency found itself in a period of "retrenchment."[34]

Unaware, of course, of the lean years to come, Krieger wrote a letter in July 1965 to Olin E. Teague, Chairman of the House Subcommittee on NASA Oversight, in response to a Congressional staff investigation into, "Future National Space Objectives." The letter set out Krieger's view of Wallops' role in the U.S. space effort and his hopes for the future, as well as providing a good insight into not only the changed nature of the research at Wallops but also into the appearance Station management wished to assume before Congress. From this perspective, what Krieger does not mention is as interesting as what is cited.[35]

He began by providing Teague with a synopsis of Wallops' background which explains that, "Never having been involved in missile development, or in very large projects where operational aspects overshadowed scientific objectives, the station's 20-year experience has been in experimental, exploratory flight testing. We have, therefore, developed a viewpoint and operating philosophy which is, perhaps, more closely akin to a laboratory than a missile range."[36]

Wallops, as has been shown, was partly established for the purpose of testing early missile designs, and the station played a major role in Langley's Scout development program and conducted many tests relating to ICBM development. Additionally, by 1965 the various meteorological programs conducted from the base had reached a point where the "operational aspects" at least equaled "scientific objectives."[37] Krieger's view of the facility remained one of a service center; and in an era when the lines between research, development, and operations, were sometimes fuzzy within NASA, he wanted Congress to have no doubts about Wallops. The Station's role consisted of assisting research "customers" in obtaining data in many areas, not just simply firing rockets or tinkering with hardware.[38]

Krieger next spelled out seven "unique capabilities" that advertised Wallops' ability to work with, coordinate, and service a wide range of users and contribute to various research agendas in a timely and economical fashion. He then explained the "needs of the scientific community," both foreign and domestic, for continued and improved sounding rocket programs. He concluded that, "Certainly there will be no decrease in the workload involved with scientific sounding rockets, and almost surely there will be a significant increase in this activity."[39] He predicted an "increase [in] the emphasis on the university Explorer class satellites," a coming need to assist "international groups in their second and third generation experiments," and "increased sophistication and technological complexity," in the payloads sent to Wallops for launch. "Wallops Station will, of course, make every effort to absorb the increased workload. ... It would be unrealistic, however, to believe that the total increase ... can be absorbed without a gradual and orderly growth in the station."[40]

Brevity was a watchword in communications of this type; certainly Krieger could not provide a detailed manifest of all Wallops' projects. What he chose to cite in his limited space were the programs he felt most apt to bring funding to the base, while keeping within established policy. Aeronautical engineering projects would not, by definition, be included in a letter dealing with "space objectives," even if Congress and NASA held them to be high priority. Krieger does not, however, talk about aerodynamic research as it might apply to space vehicle development, the kind of research performed at the base in support of Project Mercury. He writes little about engineering development work of any kind, save a brief mention of "the development," and "flight qualification," of devices for "large orbiting laboratories," via

113

sounding rockets. He says nothing about military space research, noting only that Wallops could call upon the DOD for support, and that they had conducted research "for the more scientific arms of the Department of Defense." He quite deliberately focused on space science research as practiced by universities, international partners, and (by implication) NASA.[41]

By concentrating on the needs of organizations outside of direct federal control, he emphasized the broad base of applicability and usefulness of Wallops' facilities, while making cuts in Station funding seem more a policy and less an economic decision. This also prominently displayed the civilian nature of Wallops, one of NASA's prime reasons for maintaining the base. Interestingly, Krieger did not feel it necessary to justify the basic concept of space science research by invoking specific applications or in the general terms of advancing human knowledge. He apparently took for granted that such efforts had Congressional backing as being in the national interest and enjoyed public support as well. He proceeded from a premise that space science research was both important and necessary and put forth arguments for increased funding, not a plea for institutional survival. It would appear that Wallops was perceived to be in no immediate danger of closure, only hampered by an insufficient rate of growth.

While it might be interesting to compare this 1965 view with one from an earlier era, recall that before mid-1959 Wallops existed as an extension of Langley; therefore, funding and planning depended upon the aeronautical laboratory's programs and budget. During the early 1960's when, "only a blundering fool could go up to the Hill and come back with a result detrimental to the agency," the entire space effort proceeded in such a state of fluidity that coherent long-range planning was next to impossible.[42] "Future space objectives" meant Mercury, Apollo, and the first generation applications satellites, not to mention boosters that wouldn't blow-up with such disconcerting frequency. Added to these generalities, the specifics at Wallops of the expansion of the base, its newly independent status, and the influx of new range users made it necessary for the situation to settle somewhat before planning for future programs could rationally proceed.

The letter to Teague, and the master planning episode provided the Wallops staff with their first real opportunity to look beyond the immediate program needs of an approaching fiscal year. Until this time political and technological developments dictated the nature and pace of operations at Wallops, and planning centered on specific projects or customers.[43] Now, with the initial surge giving way to a more steady effort, thought could be given to directing the flow of the program, rather than just hanging on for the ride. Most projects and programs continued to come to the base from outside sources, yet Wallops began to make some grants and sponsor programs. Though not often in the driver's seat, Wallops settled into a secure nook "on the bandwagon."[44]

The declining NASA budget affected the Station's operations, but not as drastically as at some other installations. Since much of the scale-back

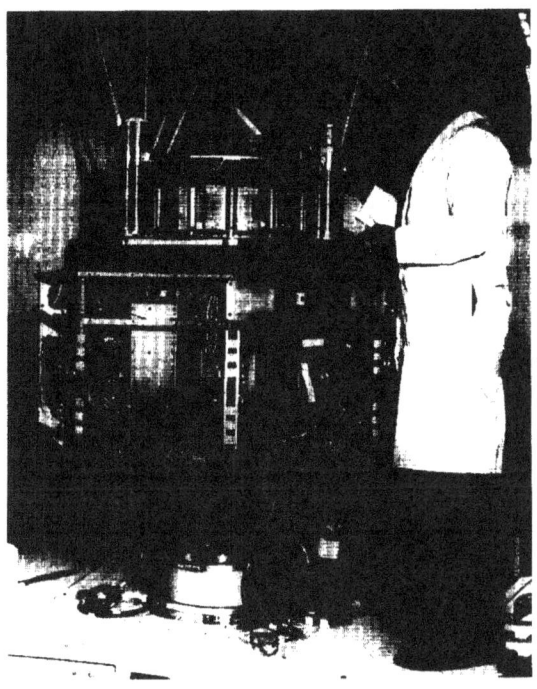

The 1970's marked the expansion of oceanic studies, including the management of the Geodynamics Experimental Ocean Satellite (GEOS 3).

involved the completion of Apollo, and Wallops had little direct stake in the piloted program, "we never had that fluctuation, it was just sort of steady."[45] The size of the civil service workforce fell by approximately 100 positions, and operational funding stabilized. (See appendix 5)[46] Construction funding at the base reflected a trend toward maintenance of existing facilities rather than the construction of new ones, and research and development funding remained stable as well.[47] The fact that Wallops provided access to space for a large number of organizations provided it with allies to fight the budget cutters. "That's ... one of the advantages they have; so many universities, so many military organizations, the foreign countries use it, and you've got all these people saying, 'oh, you can't close Wallops, we've got to have it,' ...."[48] This diffuse research program also provided no single, big-ticket line items to attract Congressional attention.

Some cutbacks did occur. *Range Recoverer* was "eventually" retired due to both economic factors and declining need for its capabilities. Scout launches from Wallops dropped by 50%. For the most part, however, the pace and level of support of the projects brought to the Station stabilized and generally continued the pattern established during the transition era.[49] Projects, either from or of interest to, the military continued to fly from the base. The expressed civilian nature of the Station kept these projects low-key but did not result from their exclusion from the launch schedule. The relative isolation of the base, favorable funding arrangements, and the fact that, despite the open atmosphere of the facility, the press and public paid little attention to the activities there served to keep the services active at the base.[50]

Space science projects from universities, international partners, NASA, and other federal agencies remained Wallops' primary stock-in-trade, however. Amid cooperation with Goddard, Langley, and other sponsors, Wallops provided project leadership for some ventures like the *Explorer 44* satellite.[51] A program of oceanographic research commenced in cooperation with university scientists who took advantage of Wallops' location on the Virginia shore to study interactions between land, marsh, and sea. In the same vein, Wallops managed the Geodynamics Experimental Ocean Satellite, *GEOS 3*, in 1975.[52] While Goddard managed the International Cooperative Program, a large portion of the foreign visitors continued to come to Wallops. Even the Soviets visited the base in 1977 and conducted research by coordinating launches from their ship, stationed off-shore, and the Station.[53] The specific experiments changed, but the overall pattern of research at the base continued.

Another area which saw change amid deeper continuity concerned Wallops' relation to its surrounding communities. Initial local grumbling over the expansion of the base settled down after the integration of the disestablished naval base into the NASA operation. Even though growth never reached the levels that some community leaders had hoped for, by 1965, "NASA [became] one of the biggest employers."[54] Aside from the transient researchers, who came to the island only for the duration of their projects, the permanent staff settled in the general vicinity of the base. Even before the creation of NASA they began to take an active interest in local affairs. In 1958, Albert P. Kellam, of the Flight Test Division, requested permission "to run and if elected, serve as councilman for the Town of Wachapreague." The election being non-partisan, Kellam was allowed to participate.[55]

Joseph Robbins indicated that this kind of civic activity was encouraged by the leadership of the Station."This was a way to get in with the community, to let the community know that we were a part of them. ... One fellow was mayor of a town, oh I guess, ten or twelve years."[56] He also related that the town of Chincoteague was allowed to make use of the excess capacity of the sewage treatment plant on the old Navy base for their own waste disposal needs. In 1965 the Virginia Bureau of Public Roads came to the Station to shoot a "driver's education film on hydroplaning."[57] Community interactions like these, in addition to the open houses and the leading role the Station played during the Ash Wednesday Storm, indicate a conscious effort to fit into the local environment. This effort did not stem from any ulterior motive so much as it arose as a consequence of the residential status of the staff. They lived in the area, so they were dealing with friends and neighbors.

116

Of course, being an active rocket range meant that Wallops could not always remain quietly in the background. Research requirements sometimes dictated launches at odd hours, or infrequent intervals. "Which was hard for the local people to understand, that you couldn't launch anything with the wind blowing a certain [way], or you couldn't launch anything if there were any ships out there you might hit, and this type of thing. ... You're sitting back in you're home, 'well boy, they're just goofing-off up there today.'"[58] In January 1965 Wallops personnel began to seek permission to gain access to nearby property in order to set up acoustic sensors needed for a project.[59] The aircraft research, which included supersonic flights offshore, and the use of guns to launch probes, also caused occasional difficulties. Despite these side-effects, "We had a good relationship, I think. There were a few farmers that blamed us for the weather conditions, you know, there are always a few superstitious people, ..."[60] The openness of the base, with its grandstands and unobtrusive security, contributed to this generally good rapport.

The community did not react to the base with quite the same exuberance shown by other communities that played host to NASA facilities. The residents of Hampton, Virginia, for example, renamed their main highway "Mercury Boulevard" and local bridges bore the names of astronauts. Quite a few retail establishments around Cape Canaveral took names with space program connotations. One does not see this around the Accomack area, though. The roads, villages, and geographic features all seem to retain the names originally bestowed upon them. Names influenced by Native American terms or religious references occur often. Businesses use commonplace names unrelated to space and rocketry. An attempt to rename the island after local native Hugh Dryden, fell through when, "a lot of people got upset, ... Wallops Island had been known as Wallops Island since the 1600's."[61] Even today, if one does not pay attention, it's easy to miss the road that leads to the base. This casual acceptance of the Station probably stems partially from the relative lack of glamorous projects (like the piloted space flight projects), and partially from the routine nature of the operations. Wallops has become just another part of the local scene.

A major concern of Krieger's was the matter of education in support of the space program. During NASA's fifth semiannual management conference in 1961, he participated in a working group study on "Improving NASA's Weight Lifting Capability," in which he commented on the problem of "obtaining the type and depth of engineering evaluation needed," for the task under study. As a part of the solution he recommended, "a strong education program by NASA, sponsoring of courses in universities, sponsoring the preparations of textbooks."[62] Krieger had a two-fold problem on his hands. First, he needed to attract and retain quality engineers, professionals who would expect the opportunity to further their education. Secondly, he noted that, "Many jobs at this Station ... appear to be somewhat

117

beyond the technical capability normally expected of a mechanic or craftsman, however skilled or dedicated. On the other hand, such a job would not seem a very challenging one for a good engineer, ..."[63]

"Bob and I conducted surveys of ... practically the whole Delmarva Peninsula. What are the needs of the people? We knew what our needs were."[64] The results of these surveys indicated that the area needed a boost in higher education: engineering for the Wallops base, agricultural for the local residents. To help meet these needs, the Station leaders sought to convince Virginia authorities to locate a branch college on the Eastern Shore. "We were courted by both the University of Virginia and VPI to request that we be made a branch of them. Well, we had no say in it, but we went up to Richmond and appeared before the Council of Higher Education, and we were assigned to the University of Virginia."[65] The branch college, originally situated in "surplus" housing just outside the base, provided services to both the local communities and to Wallops.[66] "They would fly in professors, and we had started our own technician training courses on the base, ..., rather than having our own electrical engineers teaching the courses, we contracted over with this new 'branch,' ..., to do all this."[67] Abraham Spinak recognized the importance of this concern for education by noting, "We kept our engineers that way."[68]

The residents of the area also profited. Course offerings not only included technical subjects like Trigonometry and "Advanced Engineering Math II" but also liberal arts classes such as American History and "Principles of Organization and Management," and "general studies" such as Art and "Basic Grammar Review."[69] This type of program almost certainly would not have come to the Eastern Shore during the early 1960's if not for the efforts of Krieger and the staff at Wallops. Given that university students with small projects comprised one of the new groups of customers coming to the base, and Krieger's conviction that, "If the national space program is to capitalize on this resource ..., one must think in terms of experiments that can be performed by Ph.D. candidates," the interest in higher education locally meshed well with Wallops' activities.[70]

One such activity entailed a program in August 1965 sponsored by NASA and the University of Virginia, that brought 32 biologists to Wallops. A three week course was conducted, designed to teach "operational and engineering aspects of space flight."[71] Biological payloads were not new to the base; the biggest press draw had been the flights of the monkeys (which had included "insect eggs, larvae, bacteria cultures, and cell tissue") during Project Mercury.[72] The current program incorporated launches of white rats in order to train the biologists in investigative techniques pertaining to researching the impact of space flight on living organisms.[73]

Wallops maintained its reputation for being a versatile research facility. While the biologists practiced with rats, the Johns Hopkins Applied Physics Laboratory came to the base to investigate "clear air turbulence" for NASA and the Air Force. Satellites, meteorological soundings, military tests, balloons, and "gun probes," all launched from the island in 1965, 418 in all. Add to that 7 flights from Pt. Barrow and 80 from the *Croatan*, and its easy to see that Wallops kept up a fast pace.[74] The failure rate for 1965, where a malfunction caused little or no data to be gathered, was around 13%, nominal and tolerable for low-budget, unpiloted experiments.[75] This proved to be a valuable part of operations from Wallops. With so many flights each year, if one failed, another could usually be arranged. The sounding rockets had minimal backup systems in order to keep both weights and costs low. The inexpensive and unspectacular nature of the program made it easy for NASA to continue the NACA's attitude toward failure and look upon these misfortunes as learning experiences.[76]

The "unique" character of the operations at Wallops, the interactions with the local communities, and the pastoral location all served to foster a definite "esprit de corps" at the Station.[77] The old Langley methodology, as brought to the base by Krieger and the other NACA veterans, set the tone for the environment there. Informal ("anybody, from the lowest laborer, could walk into [Krieger's] office and talk to him"), independent ("we didn't get permission, we just did it, period. I'm one of those people, you do it, and you tell [Headquarters] what you're doing, and tell them the attributes, ..."), and committed to the task at hand, this atmosphere prevailed at least until the late 1970's.[78] At that time the NACA veterans began to retire; NASA "bureaucratized," and economics finally brought about the absorption of Wallops by Goddard.[79] Within a few years people who had thought of themselves as NACA/Langley personnel came to view themselves as NASA/Wallops personnel. Whether a similar shift took place in the early 1980's is a subject open to question.

The underlying theme in this important period of Wallops' history, therefore, is change amid constancy. The variety and scope of the experiments conducted at the base, the physical size and economic investment there, and its relations with the public, press, and scientific community all changed markedly during the 1957-1965 time frame. During that same era, however, the general nature of the research tools used, the Station's role in the larger organization of which it was a part, and the "operational philosophy" and methodology prevalent remained consistent. The most significant changes involved the shift from primarily aeronautical to a more varied research program with a heavy emphasis on space science investigations and the diversification of the customer base. The stability of the staff, equipment, and methodology provided a foundation of experience upon which the new programs could be built. Both of these factors allowed Wallops to stay useful to a range of programs and researchers that might otherwise have been overlooked due to the relative size and mundane nature of their experiments. Thus, Wallops, despite its small stature and uncelebrated role, contributed significantly to the early U.S. space effort.

NOTES

1. The growth of the base is dramatically illustrated by looking at the figures relating to total plant value. FY58: $3,661,000; FY63: $24,173,000; FY65: $42,978,000; FY68: $103,388,000. Figures for 58 and 68 are in *Data Book I*, 26, table 2-b. Figure for 63 is from, U.S., Congress, Senate, Committee on Aeronautical and Space Sciences, *NASA Authorization for Fiscal Year 1963, Hearings before the Senate Committee on Aeronautical and Space Sciences on H.R. 11737*, 87th Cong. 2nd sess., 6206-13S, 174. Figure for 65 is from, U.S., Congress, House, Committee on Appropriations, *Independent Offices Appropriations for 1966, Hearings before a House subcommittee of the Committee on Appropriations*, 89th Cong. 1st sess., 6502-04H, 1232.

2. Note, for example, "As NASA matured, so did the aerospace industry — in no small part due to the efforts of Glennan and James Webb, ..., to build it up." McCurdy, 167. This ultimately affected the old custom of performing work on an in-house basis, rather than contracting out.

3. "Robbins," OHI, Tape Ib: 300.

4. "Milliner," OHI, Tape Ia: 335-425, quote is near 425.

5. Rosholt, 197-227.

6. Ibid., 289-302.

7. Ibid., v.

8. Levine, 34-43.

9. Letter, Albert F. Siepert to T. Melvin Butler, 1 June 1962, in folder "Special File, May to September 62," in RGA181-l(S). "Most of you" refers to the aforementioned field administrative officials and HQ representative. See also in the same folder Butler's reply to Siepert of 14 June recommending items for the agenda.

10. Levine, 43-46. Note that this reorganization is unrelated to JFK's assassination.

11. "Robbins," OHI, Tape Ib: 275.

12. Memorandum, G. M. Truszynski for the Staff, Office of Tracking and Data Acquisition, 8 January 1962, in folder "Functions and Authority OTDA," in NASA HQ box #1. This memo lists, among other Office responsibilities, "Wallops Station Overall Operations and Budgeting." See also: "Spinak, et al.," OHI, Tape 2a: 180, Abraham Spinak: "We never worked directly for Buckley, except we did."

13. Charts, "NASA Wallops Station, Wallops Island, Virginia," March 1961, 31 August 1964, and 1 June 1967.

14. U.S., Congress, House, Committee on Science and Astronautics, *Master Planninq of NASA Installations*, H. Report 167, 89th Cong. 1st sess., 6503 15H, 7.

15. Ibid., 35.

16. Ibid.

17. Ibid., 3-6.

18. Ibid., 28. The other five facilities examined were: Jet Propulsion Laboratory, Flight Research Center, Manned Spacecraft Center, Michoud Plant, and Mississippi Test Facility.

19. Ibid., 17-18. See also: "Robbins," OHI, Tape lb: 326.

20. House Report, *Master Planning*, 29.

21. Ibid., 32.

22. Ibid., 33, emphasis in original.

23. NASA Wallops News Release, "Contract Awards During August 1965, 3 August 1965, in folder 004696 "Wallops - Contract Awards," in file tray "Wallops Flight Facility (con't)," in NHO.

24. U.S., Congress, House, Committee on Science and Astronautics, 1st *Interim Report on Master Planning of NASA Installations*, H. Report 1220, 89th Cong. 2nd sess., 6601-24H. Page 3 notes that Wallops was the first base revisited. The others were: Lewis, JPL, and the Manned Spacecraft Center.

25. Ibid., 8. Phase one of the plan mapped and inventoried the existing plant. Phase two dealt with "demolition, renovation, or new construction," concerning the phase one structures. Phase three listed plans for "future programs." Note, the staff still hoped for programs, "possibly including facilities for large launch vehicles."

26. Ibid., 14.

27. Ibid., 17. The report ends on a somewhat self congratulatory note: "There is profit in congressional attention to the development and utilization of master plans by NASA. Periodic oversight of the policy and practice should be continued by the committee."

28. U.S., Congress, House, Committee on Appropriation, *Independent Offices Appropriations for 1967, Hearings before a House subcommittee of the Committee on Appropriations*, 89th Cong. 2nd sess., 6602-OIH, part 2, 1533-4.

29. Hallion, 134. "Spinak, et al.," OHI, Tape 2b: 155. "Robbins," OHI, Tape lb: 360.

30. *Data Book I*, 11, figure 1-2. "Spinak, et al.," OHI, Tape 2b: 235. Levine, 255-6.

31. "Spinak, et al.," OHI, Tape 2b: 235; Hansen, 376n.

32. "Spinak, et al.," OHI, Tape 2b: 235; "Robbins," OHI, Tape lb: 360.

33. Levine, 202-09, 254.

34. McCurdy, 101-6.

35. U.S., Congress, House, Committee on Science and Astronautics, *Future National Space Objectives, a Staff Study of a House subcommittee of the Committee on Science and Astronautics*, 89th Cong. 2nd sess., 6607-26H, 354-56. Teague's letter went out on 29 June 1965; Krieger's reply is dated 27 July.

36. Ibid., 354.

37. Recall Krieger's complaint about the meteorological program going too soon operational in chapter 4 above. Also, "very large program" in Wallops terms seems to refer to "billion dollar satellites." Several of those interviewed used that or a similar phrase.

38. Letter, Robert L. Krieger to Floyd L. Thompson, 15 December 1961, in folder "Special File, September - December 1961," in RGA181-l(S), for Krieger's reference to "customers." This is not the only example of the term, and that fact provides insight into how the staff viewed their relationship to those who came to the base. Remember, dissatisfied customers will either complain to the management or take their business elsewhere.

39. Staff Study, *Space Objectives*, 356.

40. Ibid.

41. Ibid., 354-55.

42. McDougall, 201.

43. Memorandum, Carl A. Sandahl for Associate Director, 10 December 1962, in folder "Wallops, August 62 - February 63," in RGA181-l(C). This memo on "Visit of Wallops personnel to discuss future requirements" demonstrates one type of planning: that involving a specific customer's near-term needs. Joseph Robbins clearly states the station management's feelings toward master planning (ie.long-range planning), in "Robbins," OHI, Tape lb: 236.

44. "Spinak, et al.," OHI, Tape lb: 170. Note that the Mobile Sea Launch Expedition, the Master Planning Report, and the Teague letter all occur around 1965.

45. "Milliner," OHI, Tape lb: 300.

46. U.S., Congress, Senate, Committee on Aeronautics and Space Science, *NASA Authorization for Fiscal Year 1962*, S. Report 475 to accompany H.R. 6874, 87th Cong. 1st sess., 6106-07S, 129, notes that Project Apollo will result in 100 new positions at Wallops. Perhaps coincidentally, the number of NASA employees at the base went from 530 in FY 1964, to 420 in FY 74.Contract personnel stood at 400 in FY 63, and dropped to 209 the next year due to the relocation of the MSFN Evaluation/Training facility to Goddard. Figures are from various budget hearings cited and Marquis Academic Media, *NASA Factbook: Guide to National Aeronautics and Space Administration Programs and Activities*, 2nd ed. (Chicago: Marquis Who's Who, 1975), 327.

47. *NASA Factbook*, 324, 326; *Data Book I*, 491-2.

48. "Milliner," OHI, Tape lb: 280.

49. "Spinak, et al.," OHI, Tape 2b: 370, for *Range Recoverer*. Andrew Wilson, "Scout - NASA's Small Satellite Launcher," *Space Flight* 21, no. 11 (November 1979): 446-59. This article includes a list of the first 98 Scout launches with pertinent data. 25 of the first 50 launches occurred at Wallops (1960-67). Only 12 of the next 48 flew from that base (1967-78).

50. "Milliner," OHI, Tape la: 498, noted that Wallops has always played host to the military, "They just don't talk about it." U.S., Congress, House, Committee on Science and Astronautics, *NASA Authorization for 1962, Hearings before the House Committee on Science and Astronautics on H.R. 3238 and 6029*, 87th Cong. 1st sess., 6104-18H, part 2, 861, for an interesting exchange: Rep Bell: "Are you performing any work at Wallops Island for the Department of Defense now?" Mr. Wyatt: "Yes, sir, a great deal. We supply the facilities and the services in terms of tracking and recording of data and furnish the data then back to the user agency. *No charge.*" [Emphasis mine]. The Navy maintains a training facility for their Aegis radar system at Wallops, and (according to the 21 March 1994 issue of

Aviation Week) plans to launch targets for tests of ballistic missile interceptors from the base in late 1994.

51. *Data Book III*, 156, for *Explorer 44*. Table 3-4 on page 133 shows that NASA funding for the sounding rocket program stayed around $19 million throughout the 1970's.

52. Ibid., 343, for the GEOS, which was managed by Wallops but launched from Vandenberg on a Delta booster.

53. "Milliner," OHI, Tape la: 520 - lb: 50.

54. Ibid., Tape la: 335.

55. Memorandum, Albert P. Kellam for Personnel Officer, 25 April 1958, in folder "Special File, March - April 58," in RGA181-l(S). See also in this folder; Letter, A. G. Clement to Mrs. A. P. Kellam, 17 April 1958, and Letter, Joseph Robbins to Langley, 29 April 1958, on the same subject.

56. "Robbins," OHI, Tape la: 460.

57. Ibid., for sewage treatment plant. For film see: Memorandum, C. C. Shufflebarger for Associate Director, 28 May 1965, in folder "Special file, June - December 65," in RGA181-l(S).

58. Ibid., la: 495.

59. Memorandum, W. Latham Copeland to Albert J. Saecker, 19 January 1965, in folder "Wallops, January - June 46 [sic]," in RGA181-l(C).

60. "Milliner," OHI, Tape la: 350. See Shortal, 670-73 for problems stemming from sonic boom test flights.

61. "Milliner," OHI, Tape la: 300, for attempt to rename the island. Hansen, 391, for Hampton names which are still in use today.

62. Staff Report, "Summary of Presentations and Discussions, 5th Semi-Annual Conference, Luray Va., March 8-10 1961," 36, in box "NASA Staff Conferences," in NHO.

63. Letter, Robert L. Krieger to Distribution List, 6 December 1963, in folder "Special File, May 63 - February 64," in RGA181-l(S). See also: "Spinak, et al.," OHI, Tape 2b: 80.

64. "Robbins," OHI, Tape la: 410.

65. Ibid. VPI is Virginia Polytechnic Institute. Note that in July 1963, U.Va., VPI, and William and Mary formed the Virginia Associated Research Center in Newport News and began working under a NASA contract; *A&A, 1963*, 288.

66. "Spinak, et al.," OHI, Tape 2b: 100. "Milliner," OHI, Tape la: 423, notes that the college, reconstituted as a community college, later moved south to the town of Melfa.

67. "Robbins," OHI, Tape la: 440.

68. "Spinak, et al.," OHI, Tape 2b: 130.

69. NASA Wallops News Release #63-2, "Extension Classes Offered at Wallops," 7 January 1963, in folder 004680 "Wallops General (1958-63)," in file tray "Centers, Wallops Flight Facility," in NHO.

70. Staff Study, *Future Space Objectives*, 355, as cited in note 35 above.

71. NASA Wallops News Release #65-255, "Biologists to Begin Technology Training Program at Wallops," 8 August 1965, in folder OW-0500000-01 "Wallops Island Flight Center (NASA)," in Space History Collection, NASM.

72. Shortal, 656-57.

73. Ernest Imhoff, "Astrorats To Teach Biologists," *The Baltimore Evening Sun*, Tuesday, 13 July 1965, B1.

74. Letter, Isadore Katz to Possible Participants in Wallops Clear Air Turbulence Project, 10 August 1965, in folder "Special File, June December 65," in RGA181-1(S). Memorandum, John R. Holtz to M. W. Rosen, 30 December 1965, in folder 005064 "15.1 Sounding Rockets 1964-69," in file tray "Sounding Rockets General," in NHO. This memo lists the sounding rocket launches for the calendar years 64 and 65. Recall, however, that a number of small calibration rockets were fired in association with each research launch, thus adding to the total number of firings.

75. My figure for the failure rate is based on the launches recorded in *A&A, 1965*. 32 launches (out of 418 listed in the memo cited Ibid.) are listed in the chronology. 4 of these are listed as failures, roughly a 13% failure rate or an 87% success rate. This sampling agrees with the figures given for a "tolerable" failure rate by several sources including "Milliner," OHI, Tape 1b: 455.

76. McCurdy, 70-1, 149-55.

77. "Spinak, et al.," OHI, Tape 2a: 540.

78. "Milliner," OHI, Tape 1a: 320, for Krieger's accessibility. "Robbins," OHI, Tape 1a: 470, for independent attitude.

79. McCurdy, 1, for NASA "bureaucratization." For views on the Goddard/Wallops merger see: "Spinak, et al.," OHI, Tape 2a: 245; "Milliner," OHI, Tape 1a: 230.

APPENDIX 1

YEAR	DEVELOPMENT LAUNCHINGS	RESEARCH LAUNCHINGS
1945	4	12
1946	8	195
1947	27	241
1948	32	282
1949	18	406
1950	26	301
1951	23	231
1952	75	223
1953	30	305
1954	45	275
1955	27	219
1956	9	114

Data from, Shortal, *New Dimension*, 736-741.
Developmental launches were almost exclusively military in nature.

APPENDIX 2

All Charts from files of NASA History Office

Organization Chart: NACA (3 March 1958)

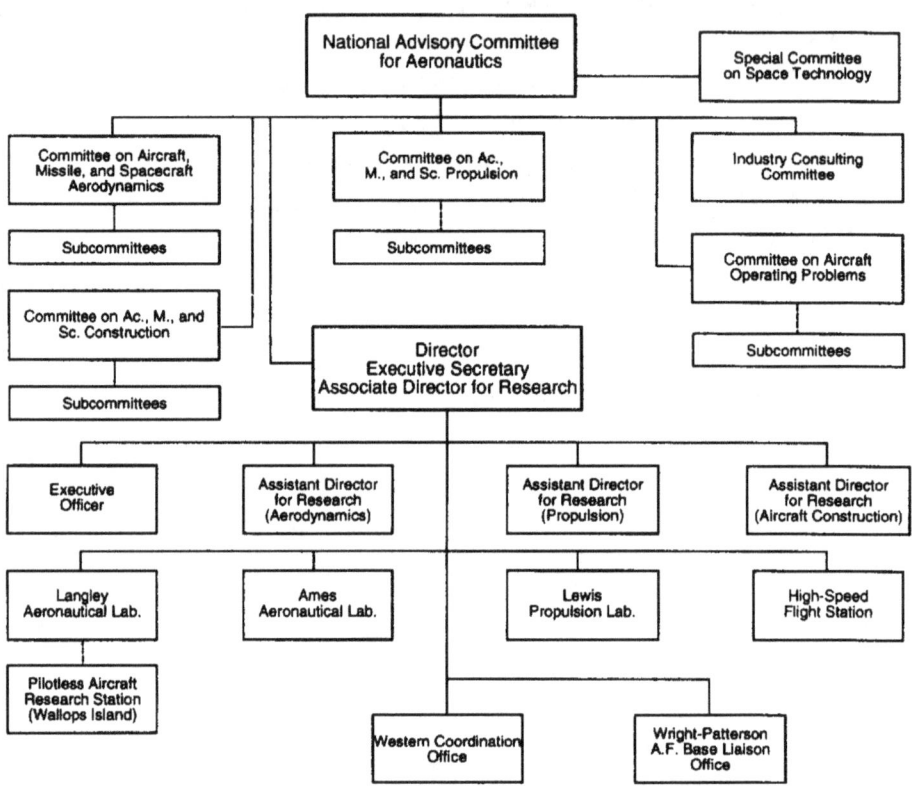

Organization Chart: NASA (21 March 1959)

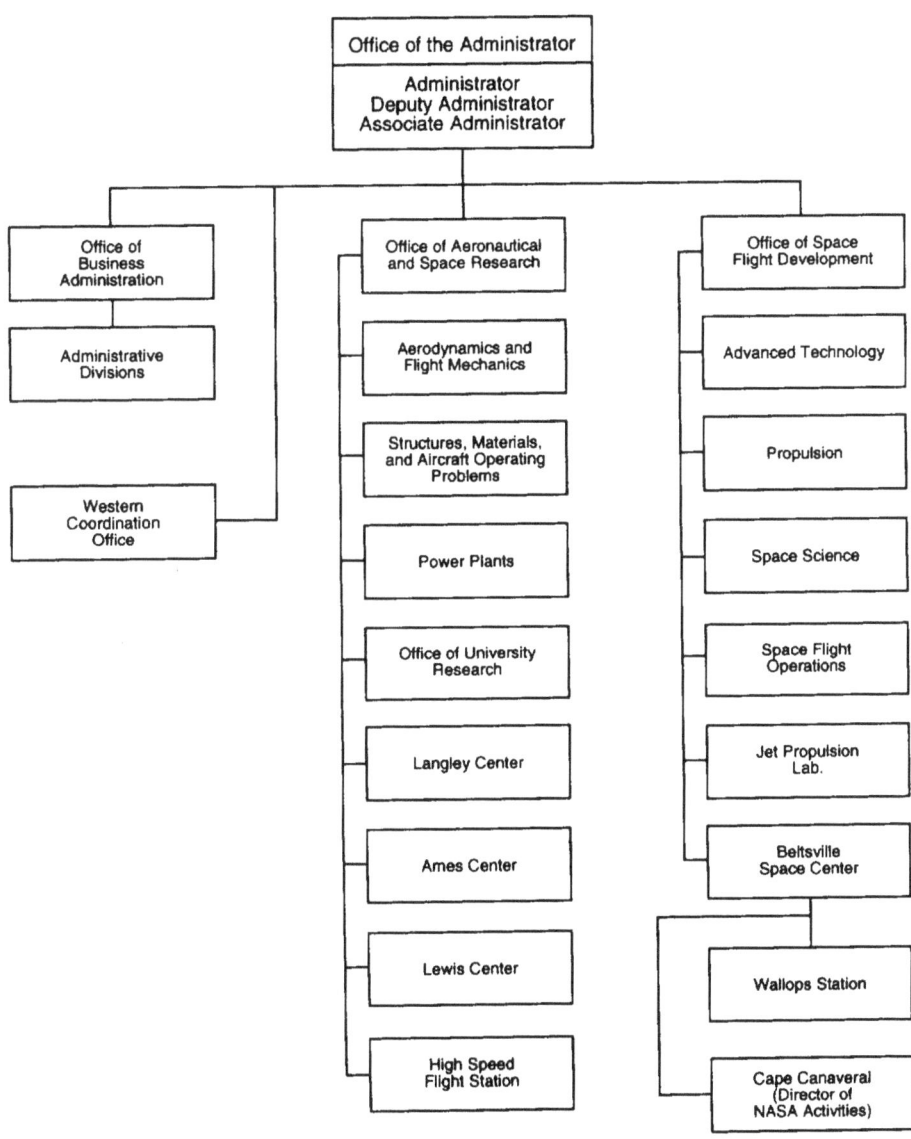

(Abbreviated for clarity)

Organization Chart: NASA (1 May 1959)

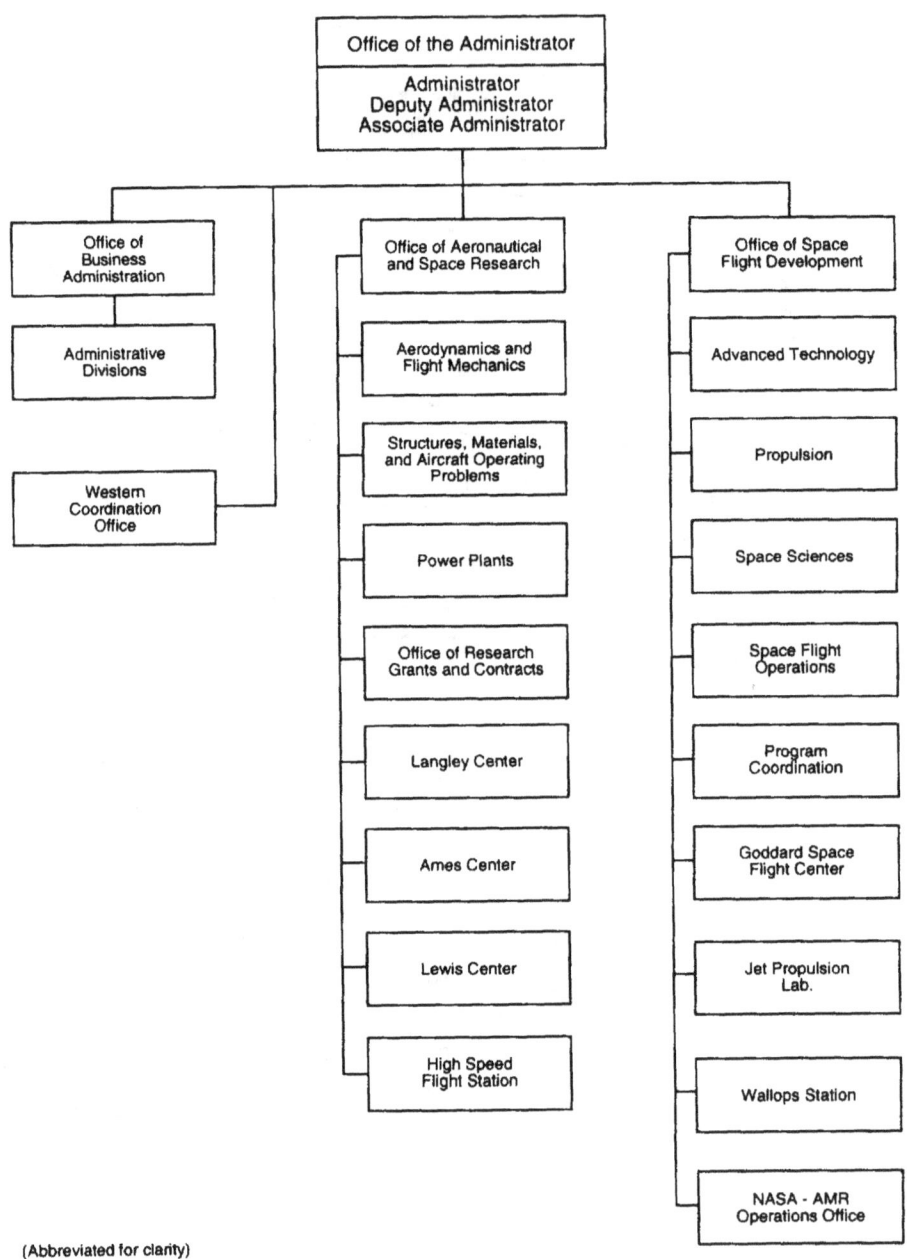

(Abbreviated for clarity)

Organization Chart: NASA (4 April 1960)

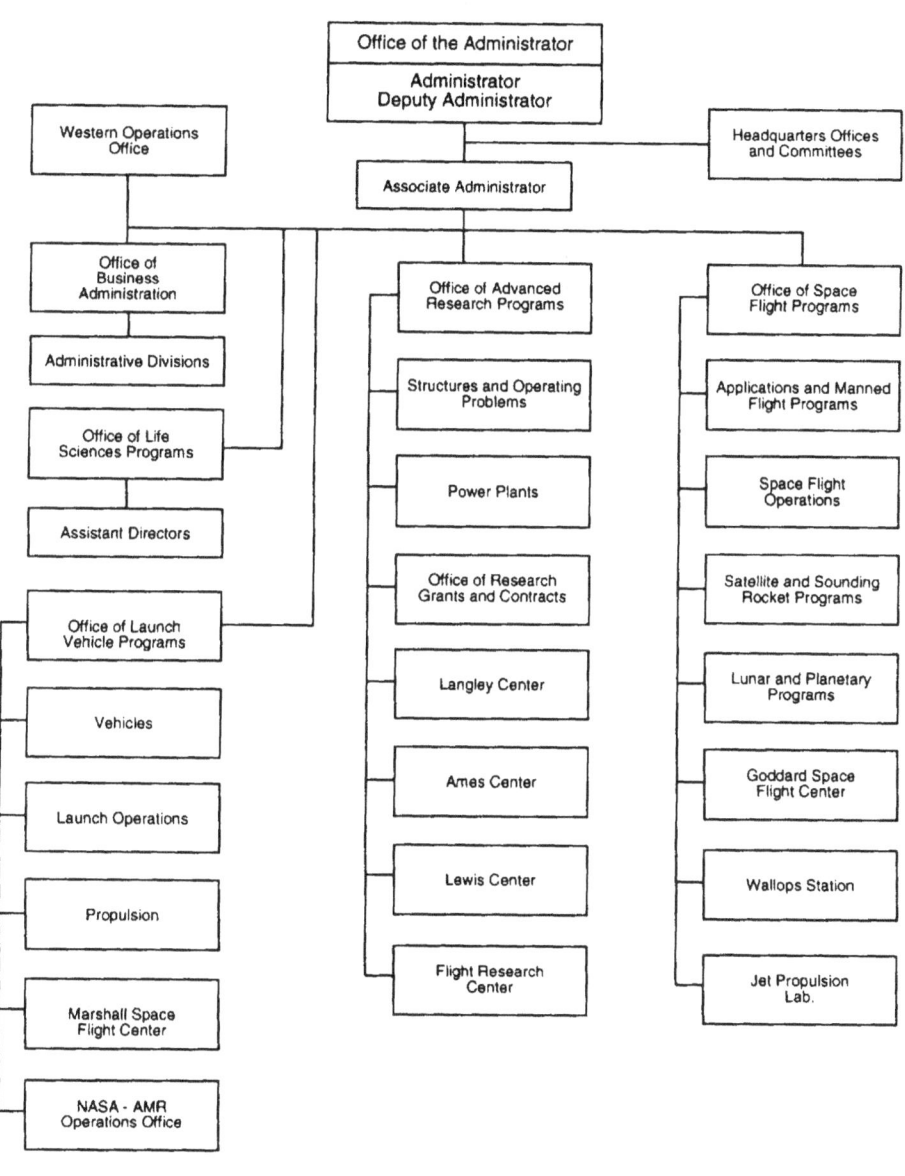

(Abbreviated for clarity)

Organization Chart: NASA (1 November 1961)

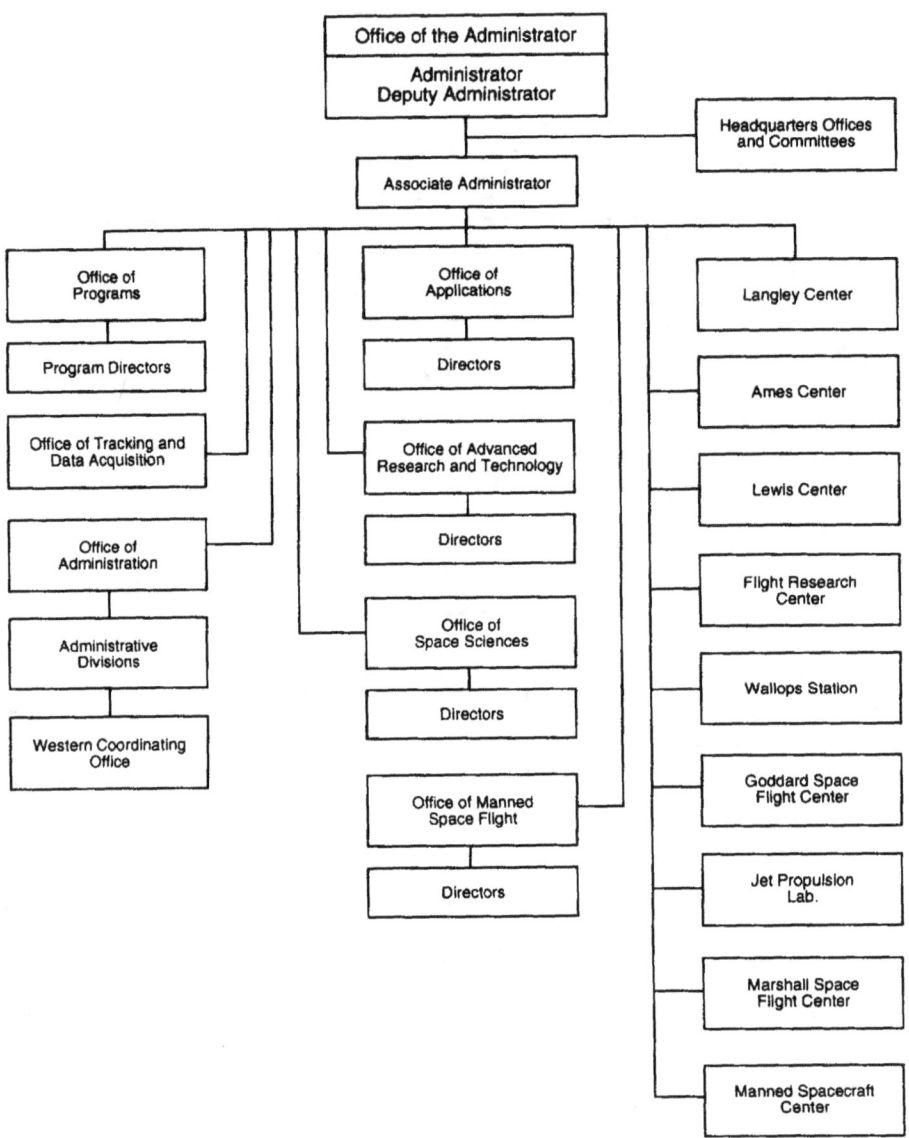

(Abbreviated for clarity)

National Aeronautics and Space Administration

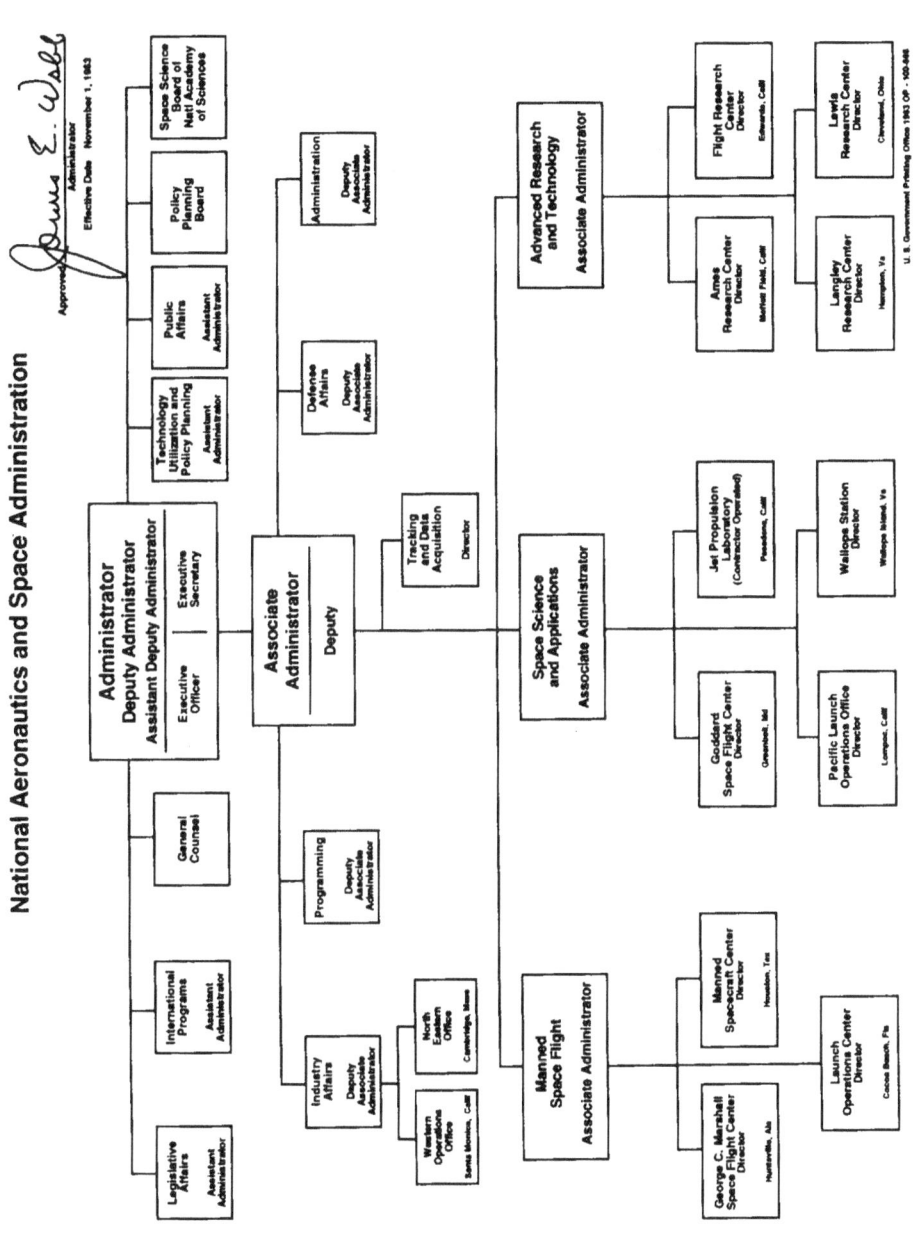

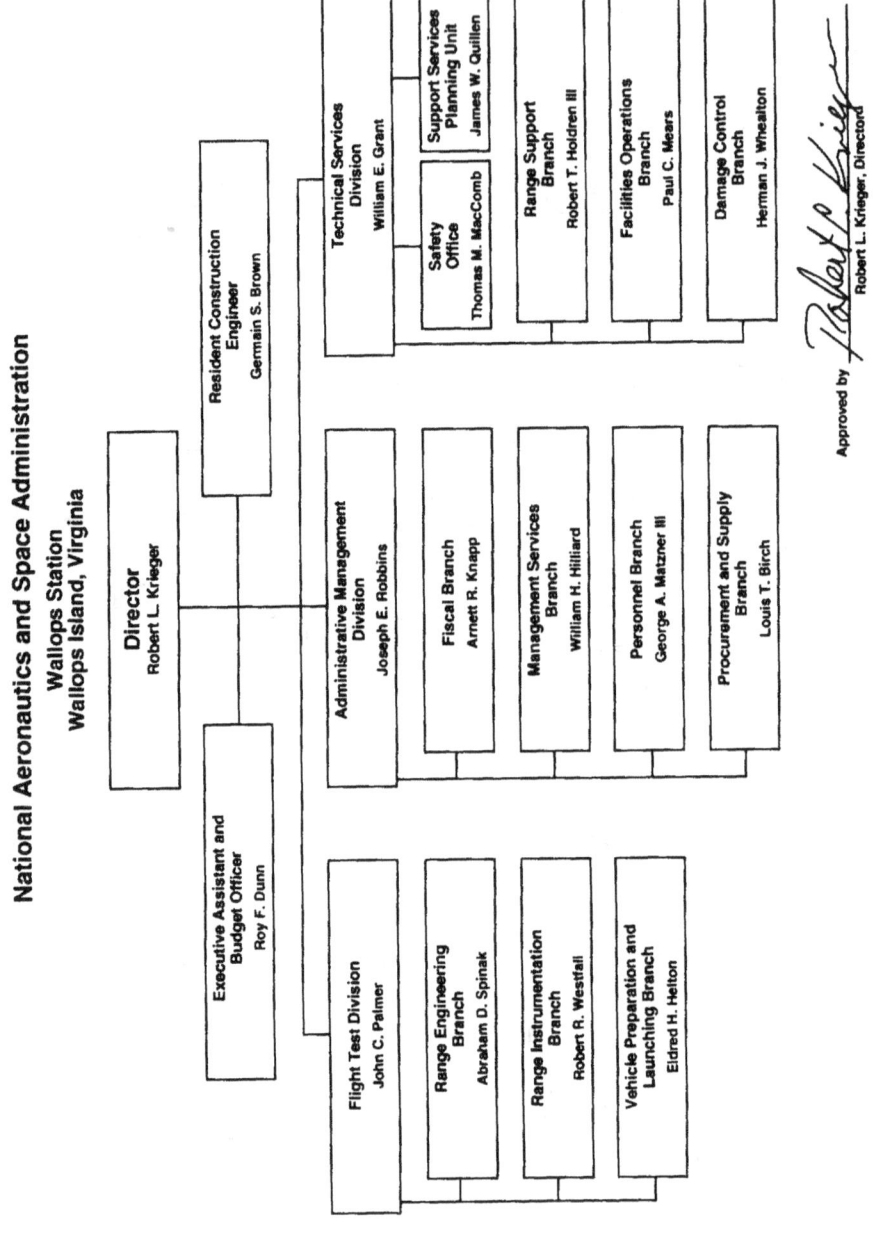

National Aeronautics and Space Administration
Wallops Station
Wallops Island, Virginia

Director
Robert L. Krieger

Resident Construction Engineer
Germain S. Brown

Executive Assistant and Budget Officer
Roy F. Dunn

Administrative Management Division
Joseph E. Robbins

Fiscal Branch
Arnett R. Knapp

Management Services Branch
William H. Hilliard

Personnel Branch
George A. Matzner III

Procurement and Supply Branch
Louis T. Birch

Technical Services Division
William E. Grant

Safety Office
Thomas M. MacComb

Support Services Planning Unit
James W. Quillen

Range Support Branch
Robert T. Holdren III

Facilities Operations Branch
Paul C. Mears

Damage Control Branch
Herman J. Whealton

Flight Test Division
John C. Palmer

Range Engineering Branch
Abraham D. Spinak

Range Instrumentation Branch
Robert R. Westfall

Vehicle Preparation and Launching Branch
Eldred H. Helton

Approved by *Robert L. Krieger*
Robert L. Krieger, Director

Date 12-18-61

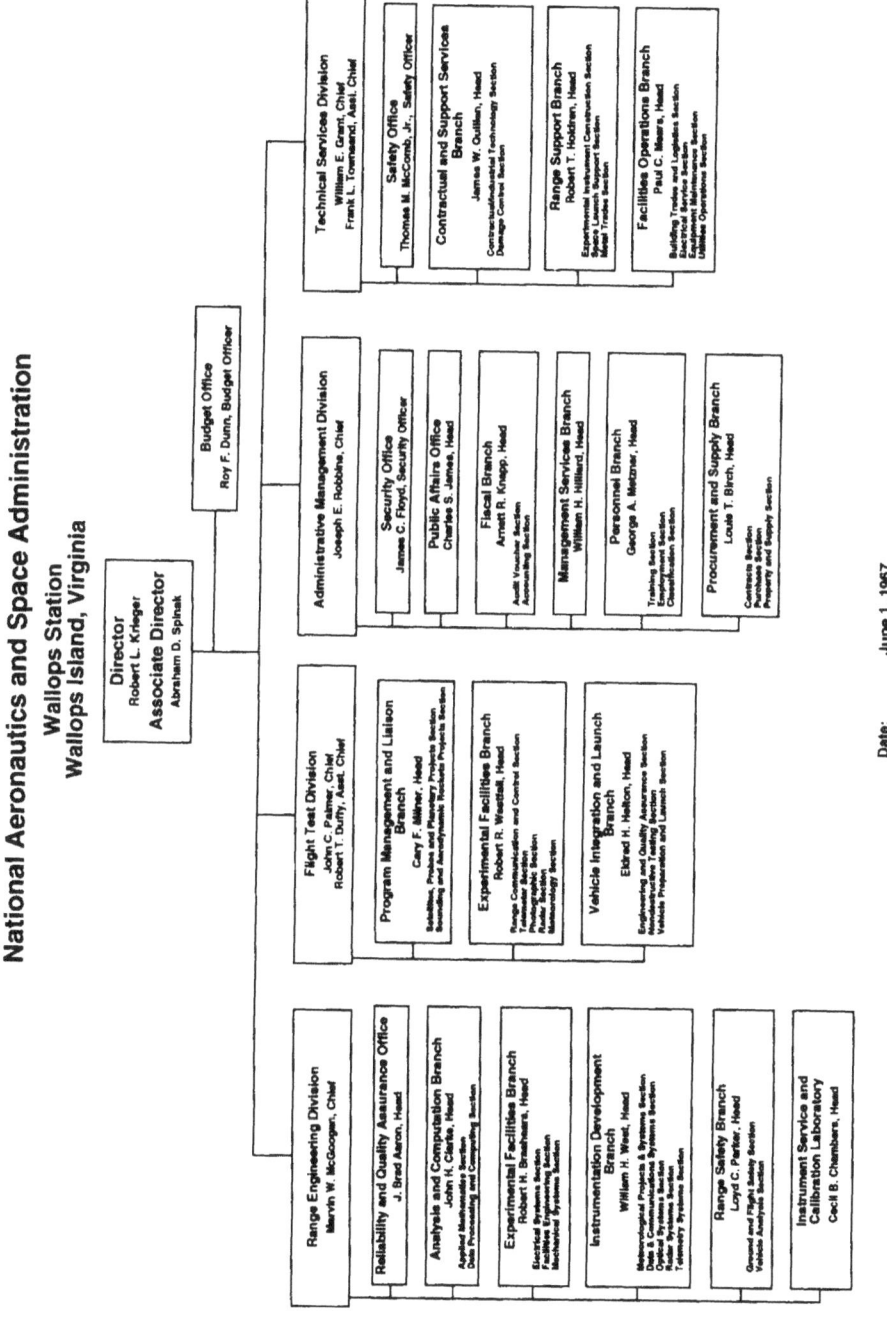

National Aeronautics and Space Administration
Wallops Station
Wallops Island, Virginia

Director
Robert L. Krieger
Associate Director
Abraham D. Spinak

Budget Office
Roy F. Dunn, Budget Officer

Technical Services Division
William E. Grant, Chief
Frank L. Townsend, Asst. Chief

Safety Office
Thomas M. McComb, Jr., Safety Officer

Contractual and Support Services Branch
James W. Guilfan, Head
Construction/Industrial Technology Section
Design Control Section

Range Support Branch
Robert T. Holdren, Head
Experimental Service Instrument Section
Space Launch Support Construction Section
Mead Trades Section

Facilities Operations Branch
Paul C. Mears, Head
Building Trades and Logistics Section
Electrical Service Section
Equipment Maintenance Section
Utilities Operations Section

Administrative Management Division
Joseph E. Robbins, Chief

Security Office
James C. Floyd, Security Officer

Public Affairs Office
Charles S. James, Head

Fiscal Branch
Arnett R. Knapp, Head
Audit Voucher Section
Accounting Section

Management Services Branch
William H. Hilliard, Head

Personnel Branch
George A. Metzner, Head
Training Section
Employment Section
Classification Section

Procurement and Supply Branch
Louie T. Birch, Head
Contracts Section
Purchase Section
Property and Supply Section

Flight Test Division
John C. Palmer, Chief
Robert T. Duffy, Asst. Chief

Program Management and Liaison Branch
Cary F. Miller, Head
Satellite, Probe and Planetary Projects Section
Sounding and Aerodynamic Vehicle Projects Section

Experimental Facilities Branch
Robert R. Westfall, Head
Range Communication and Control Section
Telemeter Section
Photographic Section
Radar Section
Meteorology Section

Vehicle Integration and Launch Branch
Eldred H. Helton, Head
Engineering and Quality Assurance Section
Nondestruction Testing Section
Vehicle Preparation and Launch Section

Range Engineering Division
Marvin W. McGoogan, Chief

Reliability and Quality Assurance Office
J. Brad Aaron, Head

Analysis and Computation Branch
John H. Clarke, Head
Applied Mathematics Section
Data Processing and Computing Section

Experimental Facilities Branch
Robert H. Braalhaars, Head
Electrical Systems Section
Facilities Engineering Section
Mechanical Systems Section

Instrumentation Development Branch
William H. West, Head
Meteorological Projects & Systems Section
Data & Communications Systems Section
Optical Systems Section
Radar Systems Section
Telemetry Systems Section

Range Safety Branch
Loyd C. Parker, Head
Ground and Flight Safety Section
Vehicle Analysis Section

Instrument Service and Calibration Laboratory
Cecil B. Chambers, Head

Date: _____ June 1, 1967

135

APPENDIX 3

YEAR	NASA EMPLOYEES	CONTRACT EMPLOYEES
1956	78	
1957	83	
1958	83	
1959	97	
1960	228	
1961	292	
1962	407	
1963	493	400
1964	530	209
1965	555	
1966	497	
1974	430	

Data from, *Data Book 1*, 498 table 6-145; various Congressional hearings; *NASA Factbook*, 327. Information on contractors is scarce.

APPENDIX 4

NATION	1ST MISSION DATE	# OF PROJECTS
Canada	12/56	3
U.K.	6/60	2
France	1/61	4
Norway	1/61	3
Italy	1/61	3
Sweden	5/61	2
Japan	8/61	3
Pakistan	9/61	3
Australia	2/62	3
Denmark	7/62	2
E S R O	11/62	1
Bermuda	12/62	2
India	1/63	2
I.N.C.O.S.P.A.R.	11/63	1
W. Germany	9/64	1

Selected International Cooperative Programs
NASA Adminstrator's Monthly Progress Report
ESRO: European Space Research Organization
INCOSPAR: International Committee on Space Research
Some projects were multi-national, all were civilian in nature and utilized a variety of equipment.

APPENDIX 5

YEAR	R. and D.	C.O.F.	A.O.
1959	0	16.14	1.36
1960	1	0	2.65
1961	2.6	2.03	4.99
1962	0.6	11.32	7.14
1963	2.7	4.16	8.9
1964	4.3	0.51	8.78
1965	6.2	1.7	11.13
1966	7.5	1.05	9.35

Wallops Funding in $ millions
R and D: Research and Development
COF: Construction of Facilities
AO: Administrative Operations
Data Book 1, 491, table 6-147.

NOTE ON SOURCES

A: Langley Research Center Historical Archives

The sources used in this thesis that were examined at Langley Research Center, Hampton, Virginia, are contained in four Record Groups. These are:

   Record Group A181-1 "Special Files," cited herein as RGA181-1 (S);

   Record Group A181-1 "Correspondence Files," cited herein as RGA181-1 (C);

   Floyd L. Thompson Papers, cited herein as FLT Papers;

   Milton Ames Collection, cited herein as MA Collection.

I was informed by the staff at Langley that RGA181-1(S), and RGA181-1 (C), were scheduled to be retired to the National Archives in late 1993. These two record groups, therefore, may no longer be located at Langley.

B: National Archives and Records Administration

There are essentially two types of NASA records at NARA, those that have been transferred to NARA's control, and those that are still under NASA's control. Both types are labeled as Record Group 255. Those under NARA control were examined, but provided little assistance with this project. Those under NASA's control can be accessed through the NASA History Office, which keeps a series of binders listing the accession forms. However, the records are organized by accession date, not the date they were generated, therefore, after examining the binders, I decided that the expenditure of time needed to peruse this rather large group could be more profitably spent on other sources.

C: National Aeronautics and Space Administration Headquarters

Several collections in the NASA History Office, Washington, DC (herein cited as NHO), provided information for this thesis. Record collections used that are contained in filing trays include:

   Administrator's Collection;

   Biographical Collection;

   Budget Materials Collection;

   Field Centers Collection;

   Program and Projects Collection.

Records contained in file boxes include:

   "NASA Headquarters Organization, OART (con't), OTDA, OSC," cited herein as NASA HQ box #1;

   "NASA Headquarters Organizations: OLV, OAST, OART," cited herein as NASA HQ box #2;

   "NASA Administrator's Monthly Progress Report," cited herein as NASA APR, (date).

The Congressional Records cited herein are all contained in the files of the NASA History Office. For the sake of clarity I have utilized the numbering system employed by that Office in my citations. This numbering system, based on the printing date of the document in question, allows one to quickly retrieve a desired record from the files, as well as saving some space in already crowded endnotes. For example: 6601-17H, indicates a House document dated 17 January 1966. NASA's Congressional Collection is organized in ascending numerical order.

D: National Air and Space Museum

Several documents for this thesis were found in the Space History Collection at the Washington, DC museum.

E: Joseph A. Shortal, *A New Dimension*

This 1978 reference tome, the only published work of substance dealing exclusively with the Wallops Station, has not generally been used as a secondary source during the course of this thesis. Shortal for many years included Wallops as part of his Pilotless Aircraft Research Division at Langley, and thus produced a work that is as much memoir and chronicle, as it is history. He participated in much of the early history of the Station (until approximately 1961), and knew the people and issues involved, so I have utilized his work (outside of the introductory chapter) in much the same fashion in which I have utilized the oral histories.

F: Oral History Interviews

I have conducted a series of four interviews with six employees of the Wallops Station. These are:

Interview #1, conducted 21 December 1993, with Abraham D. Spinak, Marvin W. McGoogan, and Robert T. Duffy, on two 90 minute cassette tapes; cited herein as "Spinak, *et al.*," OHI, Tape [number, side: tape counter reading].

Interview #2, conducted 11 January 1994, with Joseph E. Robbins, on one 90 minute cassette tape; cited herein as "Robbins," OHI, Tape [number, side: tape counter reading].

Interview#3, conducted 25 January 1994, with Joyce B. Milliner, on one 90 minute cassette tape; cited herein as "Milliner," OHI, Tape [number, side: tape counter reading].

Interview #4, conducted 19 April 1994, with James Chris Floyd, on one 90 minute cassette tape; cited herein as "Floyd," OHI, Tape (number, side: tape counter reading].

G: Wallops Flight Facility

Documentary material examined at Wallops are situated in two locations. Much general information and public relations information are contained in the office of the Public Information Officer (currently Kieth Koehler). Jack Palmer's logbooks and several radar logbooks are also in this office. The Wallops Flight Facility Records Collection (cited herein as WFFRC) yielded several boxes of general information, and 12 boxes of research material Joseph Shortal used in writing his reference volume. Box #4 of this group of 12 proved very useful and is cited herein as Wallops Box #4.

SELECTED BIBLIOGRAPHY

DOCUMENT COLLECTIONS

Hampton, VA. Langley Research Center. Floyd L. Thompson Papers.
———. Milton Ames Collection.
———. Record Group A181-1. Correspondence Files.
———. Record Group A181-1. Special Files.
Suitland, MD. National Archives and Records Administration. Record Group 255. NACA Records.
Wallops Island, VA. Wallops Flight Facility. Records Collection.
Washington, D.C. National Aeronautics and Space Administration History Office. Biography File.
———. Congressional Records Collection.
———. NASA Administrators File.
———. NASA Centers File.
———. NASA Chronologic File.
———. NASA Headquarters File.
———. Organization and Management Collection.
———. Project Files.
Washington, D.C. National Air and Space Museum. Space History Collection.

INTERVIEWS

Floyd, James C. Interview. 19 April 1994.
Milliner, Joyce B. Interview. 25 January 1994.
Robbins, Joseph E. Interview. 11 January 1994.
Spinak, Abraham D.; McGoogan, Marvin W.; and Duffy, Robert T. Interview. 21 December 1993.

BOOKS

Anderson, Frank W. Jr. *Orders of Magnitude: A History of NACA and NASA, 1915-1980*. Washington, D.C.: NASA, 1981.
Augenstein, Bruno W. *Evolution of U.S. Military Space Program, 1945-1960: Some Key Events in Study, Planning, and Program Development*. Santa Monica, CA.: Rand Corp., 1982.
Brooks, Courtney G.; Grimwood, James M.; and Swenson, Loyd S. Jr. *Chariots for Apollo: A History of Manned Lunar Spacecraft*. Washington, D.C.: NASA, 1979.
Bulkeley, Rip. *The Sputniks Crisis and Early United States Space Policy*. Bloomington: Indiana University Press, 1991.
Coletta, Paolo E., ed. *United States Navy and Marine Corps Bases, Domestic*. West Port, CT.: Greenwood Press, 1977.

Collins, Martin J., and Fries, Sylvia D., eds. *A Spacefaring Nation: Perspectives on American Space History and Policy*. Washington D.C.: Smithsonian Institution Press, 1991.

Corliss, William R. *NASA Sounding Rockets, 1958-1968: An Historical Summary*. Washington D.C.: NASA, 1971.

Davies, Merton E., and Harris, William R. *RAND's Role in the Evolution of Balloon and Satellite Observation Systems and Related U.S. Space Technology*. Santa Monica, CA.: RAND Corp., 1988.

Ertel, Ivan D., and Morse, Mary Louise. *The Apollo Spacecraft: A Chronology*. Vol 1: Through November 7, 1962. Washington D.C.: NASA, 1969.

Ezell, Linda Neuman. *NASA Historical Data Book*. Vol 2: *Programs and Projects, 1958-1968*. Washington D.C.: NASA, 1988.

_____. *NASA Historical Data Book*. Vol 3: *Programs and Projects, 1969-1978*. Washington D.C.: NASA, 1988.

Gilruth, Robert R. "From Wallops Island to Project Mercury." In *History of Rocketry and Astronautics*. American Astronautical Society History Series, Vol 7, part 2. San Diego, CA.: American Astronautical Society, 1986.

Glennan, T. Keith. *The First Years of the National Aeronautics and Space Administration: Events and Impressions as Recalled by T. Keith Glennan, 1st Administrator of NASA*. Cleveland: by the Author, 1964.

Green, Constance McL., and Lomask, Milton. *Vanguard: A History*. Washington D.C.: NASA, 1970.

Hallion, Richard P. *On The Frontier: Flight Research at Dryden, 1946-1981*. Washington D.C.: NASA, 1984.

Hansen, James R. *Engineer in Charge: A History of Langley Aeronautical Laboratory, 1917-1958*. Washington D.C.: NASA, 1987.

Hartman, Edwin P. *Adventures in Research: A History of Ames Research Center, 1940-1965*. Washington D.C.: NASA, 1970.

Hirsh, Richard, and Trento, Joseph J. *The National Aeronautics and Space Administration*. New York: Praeger, 1973.

Kash, Don E. *The Politics of Space Cooperation*. West Lafayette, IN.: Purdue Research Foundation, 1967.

Killian, James R. Jr. *Sputnik, Scientists, and Eisenhower: A Memoir of the First Special Assistant to the President for Science and Technology*. Cambridge, MA.: MIT Press, 1977.

Kistiakowski, George B. *A Scientist at the White House: The Private Diary of President Eisenhower's Special Assistant for Science and Technology*. Cambridge, MA.: Harvard University Press, 1976.

Levine, Arnold S. *Managing NASA in the Apollo Era*. Washington D.C.: NASA, 1982.

Logsdon, John M. *The Decision to Go to the Moon: Project Apollo and the National Interest*. Cambridge, MA.: MIT Press, 1970.

McCurdy, Howard F. *Inside NASA: The Changing Culture of the American Space Program*. Baltimore: Johns Hopkins University Press, 1992.

McDougall, Walter A. *The Heavens and the Earth: A Political History of the Space Age*. New York: Basic Books, 1985.

Moger, Allen W. *Virginia: Bourbonism to Byrd, 1870-1925*. Charlottesville: University Press of Virginia, 1968.

NASA. *Astronautics and Aeronautics*. Chronology. Washington D.C.: Government Printing Office, Annual, 1961-1979.

Newell, Homer E. *Beyond the Atmosphere: Early Years of Space Science*. Washington D.C.: NASA, 1980.

Neustadt, Richard E., and May, Ernest R. *Thinking in Time: The Uses of History for Decision Makers*. New York: The Free Press, 1986.

Oberg, James E. *Red Star in Orbit*. New York: Random House, 1981.

Ordway, Frederick I. III, and Sharpe, Mitchell R. *The Rocket Team*. New York: Thomas Y. Crowell, 1979.

Pisano, Dominic, and Lewis, Cathleen S., eds. *Air and Space History: An Annotated Bibliography*. New York: Garland, 1988.

Roland, Alex. *A Spacefaring People: Perspectives on Early Spaceflight*. Washington D.C.: NASA, 1985.

_____. *Model Research: The National Advisory Committee for Aeronautics, 1915-1958*. Washington D.C.: NASA, 1985.

Rosenthal, Alfred. *Venture into Space: Early Years of Goddard Space Flight Center*. Washington D.C.: NASA, 1985.

Rosholt, Robert L. *An Administrative History of NASA, 1958-1963*. Washington D.C.: NASA, 1966.

Schoettle, Enid Curtis Bok. "The Establishment of NASA." In *Knowledge and Power: Essays of Science and Government*. Edited by Sanford A. Lakoff. New York: The Free Press, 1966.

Shortal, Joseph A. *A New Dimension, Wallops Island Flight Test Range: The First Fifteen Years*. Washington D.C.: NASA, 1978.

Skoog, A. Ingemar. *History of Rocketry and Astronautics*. American Astronautical Society History Series, Vol 10. San Diego, CA.: Univelt, Inc., 1990.

Swenson, Loyd S.; Grimwood, James M.; and Alexander, Charles C. *This New Ocean: A History of Project Mercury*. Washington D.C.: NASA, 1966.

United States Navy Bureau of Yards and Docks. *Building the Navy's Bases in World War II*. Washington D.C.: Government Printing Office, 1947.

U.S. President. *Public Papers of the Presidents of the United States*. Washington D.C.: Office of the *Federal Register*, National Archives and Records Administration, 1953- . Dwight D. Eisenhower, 1960-61; John F. Kennedy, 1961.

Van Nimmen, Jane; Bruno, Leonard C.; and Rosholt, Robert L. *NASA Historical Data Book*. Vol 1: *NASA Resources 1958-1968*. Washington D.C.: NASA, 1988.

JOURNAL ARTICLES

Hetherington, Norriss S. "Winning the Initiative: NASA and the U.S. Space Science Program." *Prologue* 7 (1975): 99-107.

Koppes, Clayton B. "The Militarization of the American Space Program: An Historical Perspective." *Virginia Quarterly Review* 60 (Winter 1984): 1-20.

McDougall, Walter A. "Sputnik, the Space Race, and the Cold War." *Bulletin of the Atomic Scientists* 41 (May, 1985): 20-25.

Mack, Pamela E. "Space History." *Technology and Culture* 30 (1989): 657- 665.

INDEX

Naval Ordnance Test Station: 95
Naval Research Laboratory (NRL): 14, 23, 27, 29, 42n
Norfolk Naval Air Station: 33
Newell, Homer E. Jr.: 14, 80, 88, 109, 112
New Hampshire, University of: 95, 106n
Nike (Booster): 13, 15, 20n, 83, 92, 94
Nimbus (Satellite): 83, 84
North Atlantic Treaty Organization (NATO):
  Advisory Group for Aeronautical Research (AGARD): 16
Norway: 89, 103n

Oberth, Hermann: 51n
Office of Defense Mobilization (ODM): 42n
Offices, NASA Headquarters: *see* National Aeronautics and Space
  Administration
Ohio State University: 98n
Orbiter Project: *see* Army, U.S.
Orbiting Observatories: 90
O'Sullivan, William J.: 6, 14-15, 81, 94
Organized Labor: 64, 75n
Ostrander, Don: 38, 51n

Pacific Missile Range (PMR): *see* Vandenberg AFB
Pakistan: 89, 93, 103n
Palmer, John C. (Jack): 10, 26, 35, 45n, 110
Paraglider: *see* Rogallo Wing
Patrick Air Force Base: *see* Cape Canaveral
Pickering, William H.: 45n
Pilotless Aircraft: 5, 10
Pilotless Aircraft Research Division (PARD): *see* Langley Laboratory
Plant Security Inc.: 75n
Point Arguello, CA: *see* Vandenberg AFB
Point Barrow, AK: 92, 118
Point Mugu, CA: *see* Vandenberg AFB
Polaris (Missile): 13, 24, 41, 42n, 53n
Pre-Flight Jet: 6, 13, 18n, 41, 57, 65, 75n
President's Commission on Government Employment Policy: 34, 49n
President's Science Advisory Committee (PSAC): 23, 25, 42n
Price, Paul A.: 100n
Puerto Rico: 31
Purser, Paul: 13, 56, 71
Pyle, James T.: 32

Quesada, Elwood: 31-32

THE NASA HISTORY SERIES

Reference Works, NASA SP-4000:

Grimwood, James M. *Project Mercury: A Chronology.* (NASA SP-4001, 1963).

Grimwood, James M., and Hacker, Barton C., with Vorzimmer, Peter J. *Project Gemini Technology and Operations: A Chronology.* (NASA SP-4002, 1969).

Link, Mae Mills. *Space Medicine in Project Mercury.* (NASA SP-4003, 1965).

Astronautics and Aeronautics, 1963: Chronology of Science, Technology, and Policy. (NASA SP-4004, 1964).

Astronautics and Aeronautics, 1964: Chronology of Science, Technology, and Policy. (NASA SP-4005, 1965).

Astronautics and Aeronautics, 1965: Chronology of Science, Technology, and Policy. (NASA SP-4006, 1966).

Astronautics and Aeronautics, 1966: Chronology of Science, Technology, and Policy. (NASA SP-4007, 1967).

Astronautics and Aeronautics, 1967: Chronology of Science, Technology, and Policy. (NASA SP-4008, 1968).

Ertel, Ivan D., and Morse, Mary Louise. *The Apollo Spacecraft: A Chronology, Volume I, Through November 7, 1962.* (NASA SP-4009, 1969).

Morse, Mary Louise, and Bays, Jean Kernahan. *The Apollo Spacecraft: A Chronology, Volume II, November 8, 1962-September 30, 1964.* (NASA SP-4009, 1973).

Brooks, Courtney G., and Ertel, Ivan D. *The Apollo Spacecraft: A Chronology, Volume III, October 1, 1964-January 20, 1966.* (NASA SP-4009, 1973).

Ertel, Ivan D., and Newkirk, Roland W., with Brooks, Courtney G. *The Apollo Spacecraft: A Chronology, Volume IV, January 21, 1966-July 13, 1974.* (NASA SP-4009, 1978).

Astronautics and Aeronautics, 1968: Chronology of Science, Technology, and Policy. (NASA SP-4010, 1969).

Newkirk, Roland W., and Ertel, Ivan D., with Brooks, Courtney G. *Skylab: A Chronology.* (NASA SP-4011, 1977).

163

Van Nimmen, Jane, and Bruno, Leonard C., with Rosholt, Robert L. *NASA Historical Data Book, Volume I: NASA Resources, 1958-1968.* (NASA SP-4012, 1976, rep. ed. 1988).

Ezell, Linda Neuman. *NASA Historical Data Book, Volume II: Programs and Projects, 1958-1968.* (NASA SP-4012, 1988).

Ezell, Linda Neuman. *NASA Historical Data Book, Volume III: Programs and Projects, 1969-1978.* (NASA SP-4012, 1988).

Astronautics and Aeronautics, 1969: Chronology of Science, Technology, and Policy. (NASA SP-4014, 1970).

Astronautics and Aeronautics, 1970: Chronology of Science, Technology, and Policy. (NASA SP-4015, 1972).

Astronautics and Aeronautics, 1971: Chronology of Science, Technology, and Policy. (NASA SP-4016, 1972).

Astronautics and Aeronautics, 1972: Chronology of Science, Technology, and Policy. (NASA SP-4017, 1974).

Astronautics and Aeronautics, 1973: Chronology of Science, Technology, and Policy. (NASA SP-4018, 1975).

Astronautics and Aeronautics, 1974: Chronology of Science, Technology, and Policy. (NASA SP-4019, 1977).

Astronautics and Aeronautics, 1975: Chronology of Science, Technology, and Policy. (NASA SP-4020, 1979).

Astronautics and Aeronautics, 1976: Chronology of Science, Technology, and Policy. (NASA SP-4021, 1984).

Astronautics and Aeronautics, 1977: Chronology of Science, Technology, and Policy. (NASA SP-4022, 1986).

Astronautics and Aeronautics, 1978: Chronology of Science, Technology, and Policy. (NASA SP-4023, 1986).

Astronautics and Aeronautics, 1979-1984: Chronology of Science, Technology, and Policy. (NASA SP-4024, 1988).

Astronautics and Aeronautics, 1985: Chronology of Science, Technology, and Policy. (NASA SP-4025, 1990).

Gawdiak, Ihor Y. Compiler. *NASA Historical Data Book, Volume IV: NASA Resources, 1969-1978.* (NASA SP-4012, 1994).

Noordung, Hermann. *The Problem of Space Travel: The Rocket Motor.* Ernst Stuhlinger, and J.D. Hunley, with Jennifer Garland. Editors. (NASA SP-4026, 1995).

Astronautics and Aeronautics, 1986-1990: Chronology of Science, Technology, and Policy. (NASA SP-4027, 1997).

Management Histories, NASA SP-4100:

Rosholt, Robert L. *An Administrative History of NASA, 1958-1963.* (NASA SP-4101, 1966).

Levine, Arnold S. *Managing NASA in the Apollo Era.* (NASA SP-4102, 1982).

Roland, Alex. *Model Research: The National Advisory Committee for Aeronautics, 1915-1958.* (NASA SP-4103, 1985).

Fries, Sylvia D. *NASA Engineers and the Age of Apollo* (NASA SP-4104, 1992).

Glennan, T. Keith. *The Birth of NASA: The Diary of T. Keith Glennan,* edited by J.D. Hunley. (NASA SP-4105, 1993).

Seamans, Robert C., Jr. *Aiming at Targets: The Autobiography of Robert C. Seamans, Jr.* (NASA SP-4106, 1996).

Project Histories, NASA SP-4200:

Swenson, Loyd S., Jr., Grimwood, James M., and Alexander, Charles C. *This New Ocean: A History of Project Mercury.* (NASA SP-4201, 1966).

Green, Constance McL., and Lomask, Milton. *Vanguard: A History.* (NASA SP-4202, 1970; rep. ed. Smithsonian Institution Press, 1971).

Hacker, Barton C., and Grimwood, James M. *On Shoulders of Titans: A History of Project Gemini.* (NASA SP-4203, 1977).

Benson, Charles D. and Faherty, William Barnaby. *Moonport: A History of Apollo Launch Facilities and Operations.* (NASA SP-4204, 1978).

Brooks, Courtney G., Grimwood, James M., and Swenson, Loyd S., Jr. *Chariots for Apollo: A History of Manned Lunar Spacecraft.* (NASA SP-4205, 1979).

Bilstein, Roger E. *Stages to Saturn: A Technological History of the Apollo/Saturn Launch Vehicles.* (NASA SP-4206, 1980; paperback reprint 1996).

Compton, W. David, and Benson, Charles D. *Living and Working in Space: A History of Skylab.* (NASA SP-4208, 1983).

Ezell, Edward Clinton, and Ezell, Linda Neuman. *The Partnership: A History of the Apollo-Soyuz Test Project.* (NASA SP-4209, 1978).

Hall, R. Cargill. *Lunar Impact: A History of Project Ranger.* (NASA SP-4210, 1977).

Newell, Homer E. *Beyond the Atmosphere: Early Years of Space Science.* (NASA SP-4211, 1980).

Ezell, Edward Clinton, and Ezell, Linda Neuman. *On Mars: Exploration of the Red Planet, 1958-1978.* (NASA SP-4212, 1984).

Pitts, John A. *The Human Factor: Biomedicine in the Manned Space Program to 1980.* (NASA SP-4213, 1985).

Compton, W. David. *Where No Man Has Gone Before: A History of Apollo Lunar Exploration Missions.* (NASA SP-4214, 1989).

Naugle, John E. *First Among Equals: The Selection of NASA Space Science Experiments* (NASA SP-4215, 1991).

Wallace, Lane E. *Airborne Trailblazer: Two Decades with NASA Langley's Boeing 737 Flying Laboratory.* (NASA SP-4216, 1994).

Butrica, Andrew J. Editor. *Beyond the Ionosphere: Fifty Years of Satellite Communication.* (NASA SP-4217, 1997).

Butrica, Andrews J. *To See the Unseen: A History of Planetary Radar Astronomy.* (NASA SP-4218, 1996).

Center Histories, NASA SP-4300:

Rosenthal, Alfred. *Venture into Space: Early Years of Goddard Space Flight Center.* (NASA SP-4301, 1985).

Hartman, Edwin, P. *Adventures in Research: A History of Ames Research Center, 1940-1965.* (NASA SP-4302, 1970).

Hallion, Richard P. *On the Frontier: Flight Research at Dryden, 1946-1981.* (NASA SP-4303, 1984).

Muenger, Elizabeth A. *Searching the Horizon: A History of Ames Research Center, 1940-1976.* (NASA SP-4304, 1985).

Hansen, James R. *Engineer in Charge: A History of the Langley Aeronautical Laboratory, 1917-1958.* (NASA SP-4305, 1987).

Dawson, Virginia P. *Engines and Innovation: Lewis Laboratory and American Propulsion Technology.* (NASA SP-4306, 1991).

Dethloff, Henry C. *"Suddenly Tomorrow Came...": A History of the Johnson Space Center, 1957-1990.* (NASA SP-4307, 1993).

Hansen, James R. *Spaceflight Revolution: NASA Langley Research Center from Sputnik to Apollo* (NASA SP-4308, 1995).

Wallace, Lane E. *Flights of Discovery: An Illustrated History of the Dryden Flight Research Center.* (NASA SP-4309, 1996).

General Histories, NASA SP-4400:

Corliss, William R. *NASA Sounding Rockets, 1958-1968: A Historical Summary.* (NASA SP-4401, 1971).

Wells, Helen T., Whiteley, Susan H., and Karegeannes, Carrie. *Origins of NASA Names.* (NASA SP-4402, 1976).

Anderson, Frank W., Jr., *Orders of Magnitude: A History of NACA and NASA, 1915-1980.* (NASA SP-4403, 1981).

Sloop, John L. *Liquid Hydrogen as a Propulsion Fuel, 1945-1959.* (NASA SP-4404, 1978).

Roland, Alex. Editor. *A Spacefaring People: Perspectives on Early Spaceflight.* (NASA SP-4405, 1985).

Bilstein, Roger E. *Orders of Magnitude: A History of the NACA and NASA, 1915-1990.* (NASA SP-4406, 1989).

Logsdon, John M. General Editor. With Lear, Linda J., Warren-Findley, Jannelle, Williamson, Ray A., and Day, Dwayne A. *Exploring the Unknown: Selected Documents in the History of the U.S. Civil Space Program, Volume I: Organizing for Exploration.* (NASA SP-4407, 1995).

Logsdon, John M. General Editor. With Day, Dwayne A., and Launius, Roger D. *Exploring the Unknown: Selected Documents in the History of the U.S. Civil Space Program, Volume II: External Relationships.* (NASA SP-4407, 1996).

*

www.ingramcontent.com/pod-product-compliance
Lightning Source LLC
Chambersburg PA
CBHW081446170526
45166CB00008B/2334